New Developments and Applications in Optical Radiometry

New Developments and Applications in Optical Radiometry

Proceedings of the Second International Conference held at the National Physical Laboratory, London, England, 12–13 April 1988

Edited by N P Fox and D H Nettleton

Institute of Physics Conference Series Number 92
Institute of Physics, Bristol and Philadelphia

CODEN IPHSAC 92 1–204 (1989)

British Library Cataloguing in Publication Data

New developments and applications in optical
   radiometry
   1. Radiation. Measurement. Optical
   detectors.
   I. Fox, N.P.   II. Nettleton, D.H.
   III. Series.
   539.2′028′7

ISBN 0-85498-183-7

Library of Congress Cataloging-in-Publication Data are available

Honorary Editors
      N P Fox and D H Nettleton

Published under The Institute of Physics imprint by IOP Publishing Ltd
Techno House, Redcliffe Way, Bristol BS1 6NX, England
242 Cherry Street, Philadelphia, PA 19106, USA

Printed in Great Britain by J W Arrowsmith Ltd, Bristol

Preface

The proposal to bring together radiometrists from all over the world to participate in a specialist meeting arose originally from discussions between scientists at NPL and NBS and Dr Foukal of Atmospheric and Environmental Research, Inc. The initial aim was to present recent advances in absolute radiometry made by the two National Laboratories, but it soon became clear that there would be good support for a broad-based meeting of metrologists, meteorologists and space scientists. The first international conference on radiometry duly took place in Cambridge, USA in June 1985. The meeting was a considerable success, providing a relaxed and informal forum for discussion and it was decided then that a further meeting should be arranged after two or three years, with the National Physical Laboratory in the UK as a possible venue. This meeting was arranged and held on 12 and 13 April 1988 and the proceedings follow.

The meeting was opened by Dr Dean, the Director of the National Physical Laboratory, and his opening remarks were followed by an invited keynote address given by Dr Blevin, the Chief Standard Scientist at CSIRO and President of the Consultative Committee on Photometry and Radiometry of the CIPM. The content of the second conference was broadened from the first meeting, with increased emphasis on the applications of absolute radiometry as well as a review of the developments in absolute radiometry since the last meeting in 1985. The growing interest in optical radiation measurement was evident from the number of participants which had nearly doubled compared with the first meeting in 1985; 15 contributed papers were presented as well as 15 invited papers.

At the end of the meeting, the Director of the World Radiation Centre in Switzerland agreed to organize a third meeting of radiometrists in the autumn of 1990.

The meeting was supported by the National Physical Laboratory and Oxford Instruments Ltd.

N P Fox
D H Nettleton

Contents

Contents

Inst. Phys. Conf. Ser. No. 92
Paper presented at Int. Conf. Optical Radiometry, NPL, London, 12–13 April 1988

Optical radiometry — one hundred years after Stefan and Boltzmann

W R Blevin

CSIRO Division of Applied Physics,
National Measurement Laboratory, Lindfield, Australia 2070

ABSTRACT: About 100 years have passed since thermally based, optical radiometry got onto a sound footing with the discovery of the Stefan-Boltzmann Law and the development of the first electrical substitution radiometers. The thermal approach to radiometry is still the one most widely adopted, but in recent years new approaches exploiting photodiodes of predictable quantum efficiency and electron storage rings have become competitive. Some of the radiometric highlights of the past century are reviewed and the present state of the subject is assessed.

1. INTRODUCTION

The origins of that area of metrology which we now call 'optical radiometry' are to be found in the nineteenth century. Important contributions included the discovery of the thermoelectric effect by Seebeck in 1821 and of the photoelectric effect by Hertz in 1887, the development of thermal detectors such as bolometers and thermopiles, and the establishment of a body of physical theory for heat radiation and electromagnetic radiation in general. It is an invidious task to identify specific contributions as the starting points for modern optical radiometry. It would not be unreasonable, however, to single out the theoretical confirmation by Boltzmann in 1883 of the fourth-power-of-temperature law of thermal emission, already proposed on empirical grounds by Stefan four years earlier, and the independent development by Ångström and Kurlbaum in 1893 of the first two 'absolute' radiometers to be calibrated by substituting electrical joule heating for radiant heating. In the light of the latter two dates, this 1988 conference might well be regarded as marking the centenary of the subject of optical radiometry as we now know it.

It would be wrong to omit from this brief reference to nineteenth century contributions a mention of the intense and extensive work in the 1890s on the spectral distribution of blackbody radiation, that eventually led to Planck's announcement of his celebrated law in 1901 and the foundation of the quantum theory. A detailed history of the many contributions that led to this milestone in the history of physics is available in both the German and English languages (Kangro 1970, 1976).

Interestingly, the pioneer electrical substitution radiometers (ESRs) of 1893 were developed in connection with two different fields of application that are still at the forefront of optical radiometry to-day - meteorological radiometry and photometry. K Ångström's ESR was developed in Sweden as a compensated thermopile (Ångström 1893, 1899), and was to become the basis of an internationally recognized, reference standard pyrheliometer used over several decades for measuring the direct solar irradiance at the earth's surface. Kurlbaum's ESR was based on a bolometer (Kurlbaum 1894) and developed in Germany at the Physikalisch-Technische Reichsanstalt (PTR), which is now known as the Physikalisch-Technische Bundesanstalt (PTB) and has recently celebrated its centenary as the world's first national standards laboratory. With a colleague, Lummer, Kurlbaum had been unsuccessfully exploring the possibility of developing a radiometry-based standard for the unit of luminous intensity, to replace the Hefner amyl acetate lamp that was then the German standard of light.

Table 1.   Some areas of science and technology having an interest in optical radiometry, 1888 and 1988

1888	1988
Basic physics	Basic physics
Meteorology	Meteorology and space science
Photometry	Photometric standards
	Radiometric standards
	Defence technology
	Laser technology
	Optical fibre technology
	Synchrotron radiation

Table 1 compares the areas of science and technology that to-day have a major interest in optical radiometry with the situation of 100 years ago. Meteorological interest has broadened with the development of space technology. Photometry is now based firmly on optical radiometry. Radiometric standards have become more important in their own right, largely because of the wide range of electro-optical devices now requiring them. Twentieth century technologies based on inventions such as lasers, optical fibres and synchrotrons , and military requirements such as infrared tracking systems and night-vision devices, provide new demands. Indeed, the subject of optical radiometry has undergone a significant renaissance in recent decades. The levels of radiation encountered in the various fields of application differ markedly (Table 2). There are, nevertheless, many common features of measurement to be shared.

Table 2.   Typical levels of radiation encountered in various fields of radiometry

Field	Irradiance ($W\ m^{-2}$)	Signal for 1 cm^2 aperture (mW)
Pyrheliometry	1400	140
Radiometric incandescent standard	10	1
Photometric incandescent standard	0.3	0.03
Lasers		>1
Optical fibre communications		~1

In this paper the author describes a selection of the many contributions made to optical radiometry over the past 100 years and recollects some of the highlights of his own Laboratory's work in the field. He gives his perception of the present state of the subject and attempts to identify some of the challenges for the future. Much important work is by necessity not included, and the treatment does not claim to weight all contributions fairly in a historical sense.

2. METEOROLOGICAL AND SPACE RADIOMETRY

A major challenge in this field for 100 years has been to measure and monitor accurately the solar irradiance at the earth, a quantity that is known as the solar constant when the measurement is made outside the atmosphere and at the earth's mean distance from the sun. Until 1956 two separate, detector-based radiometric scales provided the traditional basis for such measurements, the Ångström scale based on the Swedish work already referred to above (Ångström 1899) and the Smithsonian Scale 1913 based on an ESR developed by Abbott and Fowle (1908) in the USA. The Smithsonian ESR was notable in that it used a blackened cavity as the radiation target, although the resultant scale was disseminated by a non-absolute radiometer whose target was a flat, blackened disk (Abbott 1911).

Throughout the first half of this century these two Scales continued to be used as alternative bases for meteorological radiation measurements, despite the accumulation of increasing evidence that they differed from each other by several per cent (Fröhlich 1973). It became common knowledge that the Ångström Scale gave irradiance readings that were too low and the Smithsonian Scale 1913 readings that were too high. The consensus amongst meteorologists during that period was, however, that constant systematic errors were of less importance than maintaining consistency of radiation records over long periods. Moreover, all solar radiometry during that period was undertaken at ground-level or at comparatively low attitudes, and corrections for scattering within the earth's atmosphere provided a major preoccupation in themselves.

In preparation for the International Geophysical Year 1957-58 the World Meteorological Organization introduced a compromise scale known at the 'International Pyrheliometric Scale of 1956' or 'IPS 1956' (IGY Instruction Manual 1958). The intention was that measurements on the Ångström scale were to be increased by 1.5 per cent and those on the Smithsonian Scale 1939 decreased by 2.0 per cent, although neither of these objectives appears to have been achieved in the practical realisation of IPS 1956 (Fröhlich 1973). The nett result was a scale that still differed seriously from true physical measurement in SI units.

About the mid-1960s meteorological radiometry entered a much more modern phase, stimulated in particular by the possibility of measuring the solar constant from space platforms outside the earth's atmosphere. Experience had shown, moreover, that the thermal behaviour of spacecraft departed from pre-flight predictions based on IPS 1956 measurements. A range of modern, electronically balanced, cavity ESRs were developed at the Jet Propulsion Laboratory in the USA (Kendall and Berdahl 1970; Willson 1979), the Royal Meteorological Institute in Belgium (Crommelynck 1975) and the World Radiation Center in Switzerland (Brusa and Fröhlich 1975). Solar irradiance measurements by these centres, in terms of SI units, were found to agree within a few parts in one thousand, and indicated that measurements based on IPS 1956 were about 2.2 per cent too low. Consequently a new scale, known as the World Radiometric Reference (WRR), was established in 1977 to replace IPS 1956 for radiometric measurements (WMO 1977). The uncertainty of this scale in terms of SI units is stated to be 0.3 per cent.

Improved meteorological ESRs continue to be developed and are becoming so precise and stable that they are being used unattended on space platforms to monitor the solar constant for variability (Willson 1985). There have of course been many other advances in meteorological radiometry in addition to solar measurements, but space precludes discussion of them here.

3. THERMALLY BASED RADIOMETRY IN THE NATIONAL STANDARDS LABORATORIES

It has been noted above that in Germany the PTR, now the PTB, became active in radiometry very soon after its establishment in 1887, and indeed that institution has remained at the forefront of radiometric developments ever since. Other industrially developed nations soon followed Germany's example of 1887 in establishing national standards laboratories, and much of the progress in radiometry during the twentieth century has occurred in those institutions.

Until the last two decades or so these laboratories based their radiometry almost exclusively on thermal physics, using either cavity radiators of known temperature as standard sources of calculable radiance or electrical substitution radiometers as 'absolute' detectors. For broad-band radiometry the emphasis has moved increasingly towards the ESR approach, although high-temperature cavity radiators are still widely used as a basis for spectroradiometry.

An outstanding early use of a cavity radiator as a broad-band radiation standard was that of Coblentz (1915) at the US National Bureau of Standards (NBS). He used a ceramic type blackbody at temperatures in the range 1000-1150°C to calibrate incandescent carbon-filament lamps as secondary standards, taking the value of the Stefan-Boltzmann constant σ to be 5.7×10^{-8} W m^{-2} K^{-4}. These lamps were used for some decades in the USA and elsewhere. Later, Bedford (1960) at the Canadian National Research Council (NRC) developed a radiation scale based on a large, conical cavity radiator operated over the lower temperature range of 40-150°C. A comparison with the Coblentz scale revealed certain inconsistencies in the carbon-filament lamps used, but suggested that the two scales agreed to within about 0.3%. An example of a modern high-temperature, large-area, cavity radiator is that developed at the PTB (Kaase *et al* 1984).

Since the 1930s the UK National Physical Laboratory has played an important role in the development of the ESR approach to radiometry. Its interest in this field appears to have started in the 1930s, British requirements for radiometric calibrations having previously been handled by Callendar at the Imperial College of Science, London. Callendar (1910) had based his measurements on a two-cavity compensated radiometer having the unusual feature that the electrical calibration involved Peltier thermoelectric cooling and heating rather than joule heating. Guild (1937) prior to World War II, and Gillham (1962) post-war, developed a variety of ESRs that in effect defined the state of the art at that time. Of particular value was Gillham's clear exposition of the major sources of error and uncertainty in electrical substitution radiometry, such as non-uniformity of responsivity across the receiver, non-equivalence of the radiant and electrical heating, imperfect absorption and thermal conduction by black coatings, lead heating and polarity effects, heating of the radiometer case and the imperfections of practical diaphragms. Since 1960 new ESR-based radiation scales have been developed in many national laboratories, usually involving new designs of ESRs in order to reduce the above sources of error or to make them more quantifiable. In the mid-1960s an international comparison of irradiance scales organized by NPL on behalf of the Comité Consultatif de Photométrie indicated that six of the eight scales agreed to within ±0.5 per cent (Betts and Gillham 1968).

The National Standards Laboratory (NSL) in Australia, now the National Measurement Laboratory (NML), was stimulated by Gillham's work to develop its own ESRs and irradiance standards (Blevin and Brown 1967). Improved measurements were made on the thermal and optical characteristics of a range of black coatings used on radiometer receivers (Blevin and Brown 1965, 1966) and the significant absorption of filament-lamp radiation by atmospheric water-vapour was confirmed, even for path lengths of only 1 m (Blevin and Brown 1969).

Cavity-radiometer based radiometry and ESR-based radiometry are brought together in measurements of the Stefan-Boltzmann constant σ. The present 'best value' for σ is calculated from fundamental constants and equals $(5.670\ 51 \pm 0.000\ 19) \times 10^{-8}$ W m^{-2} K^{-4} (Cohen and Taylor 1986). Many experimental measurements have been made of σ over the past 90 years, but until the 1970s the agreement with theory was very poor, the experimental result typically exceeding the calculated value by one to two per cent. The accuracy of practical radiometry was generally suspect while such a large disagreement remained. Blevin and Brown (1971), in a careful measurement of σ using a room-temperature radiometer and a cavity radiator at the gold-point, largely overcame this problem and obtained a result within 0.1 per cent of the calculated value. This project accentuated the need in precision radiometry to take extreme care in excluding stray radiation and to take into account diffraction effects, even at diaphragms of large aperture (Blevin 1970). More recently Quinn and Martin (1985) made a measurement of σ at NPL that was an order of magnitude even more precise and agreed significantly more closely with the calculated value. In this measurement the cavity radiator was operated close to the triple point of water, the only exactly known temperature, and the radiometer operated at liquid-helium temperatures. In closed systems operation of the radiometer at cryogenic temperatures can have numerous advantages, including the avoidance of lead corrections by the use of superconducting leads, increased thermal diffusivity of the receiver and negligible radiative coupling between the receiver and its surrounds.

Following the above measurement of σ, NPL developed a second cryogenic radiometer to serve as its primary radiometric standard (Martin *et al* 1985). This is claimed to be capable of measuring the power of a stabilized laser beam with an uncertainty less than 5 parts in 10^5 and is probably the most accurate radiometer available to-day.

4. NON-THERMALLY BASED RADIOMETRY

Optical radiometry has been strengthened and broadened in recent years by the development of non-thermal techniques for realising radiometric standards.

One of these advances was the discovery by Geist and colleagues of NBS that certain, commercially available, silicon photodiodes were of such perfection that their internal quantum efficiency approximated to unity over much of the visible spectrum (Geist 1979, Zalewski and Geist 1980, Geist *et al* 1980). The absolute spectral responsivities of the diodes could be measured over that region by a rather simple 'self-calibration' technique, which involved use of a bias voltage to deduce the departures of the internal quantum efficiency from unity and a measurement of the reflection loss at the surface of incidence. Diffused *pn* photodiodes were used for the earlier work and found to yield results that agreed with electrical substitution radiometers to within 1 or 2 parts in 1000, but the bias procedure required for the front surfaces of these diodes tended to cause degradation (Key *et al* 1985). Geist *et al* (1981) showed that certain induced junction *np* photodiodes obviate this difficulty by not requiring bias at the front surfaces, and Gardner and Brown (1987) recently found that self-calibration measurements with such diodes at four laser wavelengths from 458 to 633 nm agreed with the Australian ESR-based scale to within 0.1 per cent. There continues to be worldwide interest in the self-calibration technique because of its simplicity and great potential, as yet only partly realised.

A second recent and exciting advance has been the use of synchrotrons and electron storage rings as a basis for optical radiometry. The applicability of these devices to radiometry in the X-ray and vacuum ultraviolet regions has been apparent ever since Schwinger (1949) developed the basic theory of synchrotron radiation. Any thought that synchrotrons might challenge blackbody radiators as the most appropriate basis of source-based radiometry in the visible spectral region was of more recent origin, however. Key and colleagues from NPL exploited two conventional synchrotrons that were available in the UK to develop spectral standards for the air ultraviolet (Key 1970, Key and Ward 1980). They found this approach to radiometry to be promising, but concluded that significant improvements in accuracy would only be likely to be possible with the much more stable conditions in a storage ring. Work continued in a number of national laboratories having access to synchrotrons, such as NBS and the Electrotechnical Laboratory of Japan. The electron storage ring BESSY was used to particular effect by Riehle and Wende (1986) of the Berlin Institute of the PTB to serve as a primary standard of spectral irradiance over the wavelength range 400–1000 nm. It was claimed that uncertainties of 0.8 per cent were obtained in calibrating filament lamps as secondary standards of spectral irradiance and that there was scope for further significant improvement.

5. THE MERGING OF PHOTOMETRY AND RADIOMETRY

In 1979 the SI base unit for photometry, the candela, was redefined by the Sixteenth General Conference of Weights and Measures so as to be related numerically to the watt per steradian for monochromatic light of frequency 540×10^{12} hertz. As a result, physical photometry has undoubtedly become a branch of the broader field of optical radiometry. Photometric measurement differs from radiometric measurement only in that spectral weighting of the results is necessary in accordance with one of the conventional CIE luminous efficiency functions of the eye. Usually this is the $V(\lambda)$ function for photopic vision whose maximum is effectively at the frequency 540 x 10^{12} hertz. The redefinition followed some early exploratory measurements at the NPL (Preston 1963, Gillham 1964), an analysis of the advantages that would accrue (Blevin and Steiner 1974), and careful consideration by the International Committee of Weights and Measures and its relevant consultative committees.

The usual practice until now has been to perform the spectral weighting by an analogue rather than a digital procedure, the light beam to be measured being modified by a broad-band colour filter which, taken together with the detector, simulates the $V(\lambda)$ function. In the case of luminous intensity measurements on a typical standard filament lamp the radiant flux eventually incident on the radiometer receiver is likely to be as low as 30 μW, as already indicated above in Table 2. Nevertheless, this low flux can be measured with appropriate accuracy on a high-quality ESR, as demonstrated in the initial radiometric realization of the photometric units in Australia (Brown 1975). In a number of more recent realizations, however, it has been found more convenient to make a two-stage measurement, the photometric measurements being made utilizing a stable silicon photodiode whose spectral responsivity has been determined previously by comparison with an ESR (e.g. Boivin *et al* 1987, Goodman and Key 1988).

The Comité Consultatif de Photométrie et Radiométrie (CCPR) in 1985-86 held its first international comparison of luminous intensity and luminous flux measurements in terms of the redefined candela (Bonhoure 1987). This project indicated that the placing of photometry on a radiometric base had greatly increased the interest in realising photometric standards. The number of laboratories that had realised the photometric units had doubled, so that 15 laboratories participated in the luminous intensity comparison and 11 in the flux comparison. This wider participation resulted in a significant reduction of the standard deviation of the mean of the results to 0.15 per cent for both the intensity and the flux comparisons, although the agreement between the national laboratories taken individually had improved only slightly. Since the experimental methods used in the realisations were much more varied than previously, it is very probable that the mean results are more exact representations of the candela and the lumen.

The BIPM maintains groups of filament lamps as secondary standards of luminous intensity and flux, and the calibrations of these had previously been aligned with the means of earlier international comparisons in terms of the pre-1979 units based on a platinum-point blackbody. In order to align the calibrations with the means of the recent comparison, and hence with the radiometric base, it was necessary to increase the previously attributed values of luminous intensity by 1.0 per cent and to reduce the previously attributed values of luminous flux by 0.7 per cent. These adjustments indicate that, prior to 1979, there had been an inconsistency of 1.7 per cent between the realisations of the lumen and the candela.

6. CONCLUSION

Important new technologies requiring optical radiometry have appeared during the last few decades, but only brief mention can be made of them here. Laser technology has brought new measurement requirements because of the high values of radiant flux often involved and the sometimes associated characteristic of very short pulse duration. On the other hand lasers, of both the fixed frequency and tunable variety, also provide a most useful tool for radiometric measurement. Optical fibre communications is a new technology calling for reliable radiometry in the near-infrared region, particularly at the wavelengths 850, 1300 and 1550 nm. This is stimulating studies of devices such as germanium photodiodes for their suitability as transfer standards. Defence technology also present new radiometric requirements.

It has been seen that modern radiometry in the visible and nearby spectral regions can be based by choice on blackbody sources, electrical substitution radiometers, synchrotron radiation or photodiodes of predictable quantum efficiency. Some of the major recent advances involve equipment that is very costly and lacking in portability, e.g. electron storage rings and cryogenically-cooled electrical substitution radiometers. This will place a greater demand on stable, portable transfer standards, and it is in this very area that optical radiometry generally remains weak. The lack of really adequate transfer standards has been a limiting factor in many radiometric and photometric intercomparisons, and this practical problem is worthy of much more active study.

Great advances have been made in radiometry over the past 100 years, and particularly over the last 20 years or so. In some respects, however, progress remains disappointing. It seems not unreasonable to ask why are so many broad-band measurements still being made? Would it not be reasonable in this day and age to expect most measurements to be made spectrally, with an incorporated microprocessor to provide in real time whatever integrations might be required. Why is so much difficulty still found, particularly in spectroradiometry, in mixing source-based and detector-based measurements?

The continuing development of optical radiometry seems well assured because of the wide and varied range of scientists and technologists now active in the field. It is probably helpful that in the different areas of application we go our separate ways for significant periods, since this is likely to produce innovative ideas and avoid excessive conformity. It is equally important, however, that radiometrists come together at intervals to share ideas and the organisers of this conference and its predecessor are to be congratulated on their initiative.

7. REFERENCES

Abbott C G and Fowle Jr. F E 1908 *Ann. Astrophys. Obs. Smithsonian Inst.* 2 34
Abbott C G 1911 *Smithsonian Inst. Misc. Collections* 56 No. 19
Ångström K 1893 *Nova Acta Soc. Sci. Upsal. ser. 3* 16 1
Ångström K 1899 *Astrophys. J.* 9 332
Bedford R E 1960 *Can. J. Phys.* 38 1256
Betts D B and Gillham E J 1968 *Metrologia* 4 101
Blevin W R 1970 *Metrologia* 6 39
Blevin W R and Brown W J 1965 *J. Sci. Instrum.* 42 385
Blevin W R and Brown W J 1966 *Metrologia* 2 139
Blevin W R and Brown W J 1967 *Australian J. Phys.* 20 567
Blevin W R and Brown W J 1969 *Metrologia* 5 28
Blevin W R and Brown W J 1971 *Metrologia* 7 15
Boivin L P, Gaertner A A and Gignac D S 1979 *Metrologia* 24 139
Bonhoure J 1987 *Metrologia* 24 157

Brown W J 1975 *Metrologia* **11** 111
Brusa R W and Frölich C 1975 *World Radiation Center Publication 543* (Davos: WRC)
Callendar H L 1910 *Proc. Phys. Soc.* **23** 1
Coblentz W W 1915 *Bull. Bur. Std.* **11** 87
Cohen E R and Taylor B N 1986 *CODATA Bulletin* No. 63 (Oxford: Pergamon)
Crommelynck D 1975 *Institute Royal Météorologique de Belgique* Publication Series A, No. 89
Fröhlich C 1973 *Proc. Symposium on Solar Radiation* ed. W H Klein & J R Hickey (Washington: Smithsonian Inst.) pp 61–78
Gardner J L and Brown W J 1987 *Appl. Opt.* **26** 2431
Geist J 1979 *Appl. Opt.* **18** 760
Geist J, Zalewski E F and Schaefer A R 1980 *Appl. Opt.* **19** 3795
Geist J, Liang E and Schaefer A R 1981 *J. Appl. Phys.* **52** 4879
Gillham E J 1962 *Proc. Roy. Soc. (London)* **A269** 249
Gillham E J 1964 *Proc. Roy. Soc. (London)* **A278** 137
Goodman T M and Key P J 1988 *Metrologia* **25** (in press)
Guild J 1937 *Proc. Roy. Soc. (London)* **A161** 1
IGY Instruction Manual 1958 *Ann. Internat. Geophys. Year* (London: Pergamon) 5 Part VI, pp 365–466
Kaase H, Bischoff K and Metzdorf J 1984 *Licht-Forschung* 6 29
Kangro H 1970 *Vorgeschichte des Planckschen Strahlungsgesetzes* (Wiesbaden: Franz Steiner)
Kangro H 1976 *Early History of Planck's Radiation Law* (London: Taylor & Francis)
Kendall J M and Berdahl C M 1970 *Appl. Opt.* **9** 1082
Key P J 1970 *Metrologia* **6** 97
Key P J and Ward T H 1978 *Metrologia* **14** 17
Key P J, Fox N P and Rastello M L 1985 *Metrologia* **21** 81
Kurlbaum F 1894 *Ann. Phys.* **287** 591
Martin J E, Fox N P and Key P J 1985 *Metrologia* **21** 147
Preston J S 1963 *Proc. Roy. Soc. (London)* **A272** 133
Quinn T J and Martin J E 1985 *Phil. Trans. Roy. Soc. (London)* **A316** 85
Riehle F and Wende B 1986 *Metrologia* **22** 75
Schwinger J 1949 *Phys. Rev.* **75** 1912
Willson R C 1979 *Appl. Opt.* **18** 179
Willson R C 1985 *Advances in Absolute Radiometry* ed P V Foukal (Cambridge, Mass.: Atmospheric and Environmental Research Inc) pp 6–11
WMO 1977 *World Meteorological Organization: World Radiometric Reference, Resolution 10 (Executive Committee XXX)* pp 136–137
Zalewski E F and Geist J 1980 *Appl. Opt.* **19** 1214

Inst. Phys. Conf. Ser. No. 92
Paper presented at Int. Conf. Optical Radiometry, NPL, London, 12–13 April 1988

11

Laser power and energy radiometry

K. Möstl

Physikalisch-Technische Bundesanstalt, Braunschweig, Fed. Rep. Germany

ABSTRACT: The paper describes how scales for laser power and energy can be built up by starting from absolute radiometers or absolute calorimeters. Examples for primary and secondary standards are discussed.

1. INTRODUCTION

Because of the differences between lasers and conventional sources, there is a need for a special laser radiometry. The main differences are as follows:

* Radiation is monochromatic. This is on the one hand an advantage, because some corrections in absolute radiometry are wavelength dependent. On the other hand, extreme monochromaticity is connected with:
* High grade of coherence. This causes problems with all optical components where interference and speckle patterns can occur, as in windows, filters, beam splitter and diffuser plates. Discussions concerning these problems can be found in the literature (e.g. Boivin 1981 and 1982, Edwards 1969). The experience can be summarized by the following statements: If interference fringes or speckles cannot be avoided, their numbers must by means of a suitable design be made so high that the occurrence of one more or less does not significantly influence the measurement result. The practical consequence is that - if it cannot be avoided - windows must be wedged and a diffusser mounted close to the detector, the sensitive area of which must be sufficiently large to collect a high number of speckles.
* High and non-uniform irradiance. This requires a broad range of linearity, an excellent spatial uniformity and a high damage threshold of the absorbing medium.
* High radiant power levels. These require a good dissipation of the power supplied, to avoid thermal damage to the absorber and heat sensor. If the detector is equipped with a heater in order to substitute electrical power for radiant power, this device is necessarily of large volume and therefore the equivalence of both kinds of heating may be seriously interfered with.
* Polarization. If the beam does not impinge perpendicularly to the detector surface (as for example in cone absorbers) a polarization dependence of the reflectance must be taken into account. The same applies if the detector comprises a beam splitter or when such a component is part of the radiometric set-up.
* Short pulses (if dealing with a pulse laser). This means that radiant energy instead of radiant power is to be to measured. In other words, a detector is needed which has a cooling time constant large enough to integrate the radiant power.

2. BASIS OF LASER RADIOMETRY

If the above listed difficulties are observed, it is possible to start from each primary standard detector and build up a scale for the radiant power and radiant energy of lasers. Primary sources are much less suitable as a starting point, because these are not line sources but (at present) continuum radiators, and to select a quasi-monochromatic beam of known radiant power would require an extremely accurate knowledge of the spectral transmission of the set-up used for the spectral selection. The primary standard detector serving as starting point can be an absolute radiometer measuring radiant power or an absolute calorimeter measuring radiant energy. The author prefers a radiometer, because with this only corrections for steady state readings have to be determined, whereas for a calorimeter, time dependent corrections must be considered. This is a much more complicated thermodynamic problem. In section 4, an example is given which proves this statement.

Figure 1 shows a cross section of the most important parts of the PTB primary laser radiometer, which is capable of measuring laser power in the range from 1 mW to 10 W. The lower limit is set by detector and amplifier noise and the upper by the load limit of the heater. The radiation absorber is a polished hollow cone which is electro-plated with an almost specular reflecting black nickel layer. As the thickness of this layer is only of the magnitude of 1 μm, the temperature drop across it is small and therefore no thermal damage is to be expected. The electric heater is wound around the cone and bonded with epoxy cement. To reduce the nonequivalence between radiation and electrical heating, the cone is enclosed by a gold plated heat shield tube, which reflects most of the thermal radiation emitted by the heater. The heat transferred to the heat shield via heat conduction through the air flows down the tube to the ground plate where the thermopile is attached; this amount of heat therefore contributes to the output signal. The residual nonequivalence (about 5 parts in 10^4) is measured according to a well-known method (Brusa 1983, Brandt

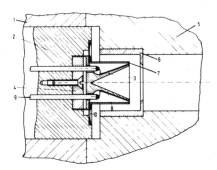

Fig. 1. Part of the absolute laser cone radiometer for power levels from 1 mW to 10 W. 1,2,4 part of the heat sink, 3 absorber cone, 6 baffle, 7 gold plated heat shield, 8 heater, 9 leads, 10 thermopile

and Möstl 1980) using the difference between the responsivities in air s_{air} and in vacuum s_{vac} for both radiative (rad) and electrical (el) heating:

$$(1) \qquad s_{rad,air}/s_{rad,vac} = f_{noneq} \cdot s_{el,air}/s_{el,vac}$$

Herein f_{noneq} is the nonequivalence correction factor. It should be mentioned that eq. (1) is only valid when the nonequivalence is exclusively produced by heat losses via air and not by effects which do not disappear in vacuum,

TABLE 1. PTB laser radiometer corrections and their uncertainties for a confidence level of 95 %

Quantity	Value	Remarks
α	0.99794 ± 0.00004	d = 5 mm; λ = 633 nm
α	0.99793 ± 0.00004	d = 5 mm; λ = 1047 nm
α	0.99745 ± 0.0012	d = 5 mm; λ = 1321 nm
f_n	1.00049 ± 0.0007	
f_l	0.99994 ± 0.0002	
f_p	0.999914 ± 0.00006	

such as thermal radiation.

Another reason for measuring the temperature increase only at the ground plate and not directly at the cone is that this improves the uniformity of response. Table 1 is a compilation of the corrections which were applied together with their 2σ uncertainties: A correction for reflection losses (α) which is almost independent of wavelength, a correction factor for the lead heating effect (f_1) and one for the nonequivalence as discussed. Finally, a small correction (f_p) must be applied if the electrical calibration is performed with only one polarity of heating current instead of both. The Peltier effect is probably responsible for this polarity dependence. The total correction factor is 0.998665 ± 0.000925 and the scale uncertainty below 0.1 %.

3. EXPANDING THE SCALE BY TRANSFER TO SECONDARY STANDARDS FOR RADIANT POWER

The next step is the transfer of the scale of the primary standard to a secondary standard. For this purpose, a laser beam is needed which is fitted to the primary standard and treated in such a way as to minimize the problems mentioned at the beginning. For example, it may be necessary to expand and homogenize the beam if a radiometer designed for conventional light sources of low power is used. The secondary standard should be chosen to exhibit the following features:

* It should be capable of measuring non-treated laser beams; this requires exellent uniformity and high damage threshold.
* It should be capable of expanding the dynamic range (a) to good levels which requires low noise performance or (b) to high power levels requiring high damage thresholds.
* It should be equipped with an electrical heater, as this is an advantage in expanding the range. It enables the user to check the detector's linearity as well as its long-term stability.

As standard and test detector are irradiated alternately, the laser must incorporate a sophisticated power stabilisation to achieve good reproducibility. Otherwise, equally good results can be obtained if the ratio technique is applied: A small portion of the laser radiation is split off by a wedged beam splitter into a reference channel to enable corrections for a drift of radiant power. It must be noticed that this method only works precisely if the reference signal is integrated with the same time constant as that of the detector which is just being irradiated by the main beam (Möstl 1988). The integrating circuit must therefore be adjustable. If these conditions are fullfilled the ratio of detector signal and reference signal can be kept constant to within 0.01 %, even when the laser power fluctuations are in the magnitude of several percents.

Figure 2 shows the cross section of a detector for high power lasers. A cavity is used to almost completely trap the radiation. The cone and the bottom plate are coated with a highly-reflecting nickel layer. The reflections on these surfaces cause an expansion of the laser beam, thus reducing the irradiance by a factor of about 30, and the blackened cylinder surface of the cavity is thus protected from exposure to extremely high levels.

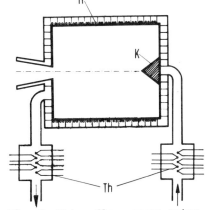

Fig. 2. Water flow power meter. H heater, K nickel-coated ccne of reflectance, Th thermopile

The cavity is cooled by a jacket of flowing water. The difference in temperature between the outflowing and the inflowing water is measured by 27 thermocouples at each side. For the electrical calibration, the instrument is equipped with an electric heater which is wound around the cavity walls. The heater is only used to test the linearity and long-term stability of the response. It is obvious that great care must be taken to keep the temperature of the inflowing water constant. Rapid changes in temperature would create a noisy output and slow changes would lead to faulty readings, as the temperature at the output cannot follow the change of input temperature instantaneously, but with a delay of about 10 s. The flowing water drains the heat and the output signal therefore depends on the water flow rate. For this reason, the flow rate v must be measured simultaneously with the thermovoltage U. This is done by means of a precision oval disk meter. The product U·v of thermovoltage and the actual flow rate is then a good choice for the detector output quantity, since it is invariant against small changes of flow rate. The calibration is performed by comparison with the cone radiometer previously described at a power level of 3 W. Unfortunately, this level is quite close to the lower limit of application of the water flow radiometer. Consequently, with a value of 0.7 %, the uncertainty of the correction factor is relatively high. An improvement would be possible if a 10 W laser were available. For the correction factor a value of 0.9913 was found:

$$(2) \qquad s_{rad} = 0.9913 \cdot s_{el}$$

In this equation the electrical responsivity s_{el} is defined by

$$(3) \qquad s_{el} = U \cdot v / P \qquad (P: \text{supplied electrical power})$$

This means if this type of power meter were applied as an absolute one without any corrections, the error would still be below 1 %. Large reading time intervals (2 min) are a disadvantage of this instrument. These are necessary because of the relatively high thermal mass of the cavity walls and the water jacket, and because of the delay of response due to the time which the water needs to run through. The occurrence of a delay time at the beginning, which was also found in other laser power meters (Möstl 1978), must be taken into account if the response time is to be improved by electronic feedback.

4. A STANDARD CALORIMETER

Figure 3 shows the cross section of a standard calorimeter which is suitable as a basis for an energy scale (Möstl et al. 1984). A brief description will demonstrate how much more complicated the corrections are for dynamic measurements than for steady-state readings. The absorbing element is a hollow cone. Unfortunately, it is not possible to blacken its surface: The high irradiance levels during the laser pulses would immediately destroy it. Metals with high melt-

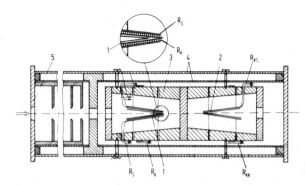

Fig. 3. Cross section of the compensated cone calorimeter; 1 measurement cone, 2 compensating cone, 3 isothermal jacket, 4 double-walled housing, 5 aperture system, R_S sensor, R_H heater, R_K, R_{KS}, and R_{KK} compensating resistors.

ing temperature are therefore used. For this instrument a chromium layer has been chosen because of its good chemical stability. As its reflectance is high many reflections are necessary to absorb almost all the radiation, and for this reason the cone angle is only 12°. Nevertheless, there is still a difference in reflectance for parallel and perpendicularly polarized radiation if the beam is not carefully centered with reference to the cone axis. Another disadvantage of the high reflectance is that it leads to a concentration of the impinging radiation in the apex region, which significantly lowers the damage threshold. Cone calorimeters can therefore only be applied in a small energy range from about 10 mJ to 100 mJ. The lower limit is set by drift and noise problems.

The calorimeter must measure energy and not power, which makes it necessary to isolate the cone thermally. For this reason the cone is held in the centre of an isothermal jacket by only thin steel wires with negligible heat conduction. To measure the temperature increase, the cone is enclosed by a bolometer winding. Below this winding there is yet another for the purpose of electrical calibration by means of current pulses. To reduce the drift problems, another identical cone is installed in a second jacket. The bolometer windings of the two cones form one branch of a Wheatstone bridge. The other one is formed by two more windings wrapped around the jackets. When an electrical calibration is performed, the heating current and the voltage drop measurements must be time resolved, as the resistance of the heater changes slightly during the heating phase due to the increasing temperature. The electrical energy supplied to the cone can then be calculated by a time integration of the product of heating current and voltage.

A very simple model of the transient response of the calorimeter should facilitate a better understanding of the experimental results: Measurements of the temperature dependence revealed that the relation between temperature and sensor resistance R_s is very closely linear (the second-order temperature coefficient is only 2.6 parts in 10^5 of the first order one). The relation between temperature rise ΔT of the cone and the change of the sensor resistance ΔR is therefore given by:

(4) $\qquad \Delta R = a \cdot R_s \cdot \Delta T \qquad (a = 3.88 \times 10^{-3} \; K^{-1})$

In the second experiment in which the heater was charged with constant electrical power instead of pulses, the steady-state temperature rise was determined from resistance measurements by using eq.(4). The result was a linear relation between the electrical power P and the temperature rise. This means that the heat transfer from the cone to the surroundings can be described by a constant heat resistance W:

(5) $\qquad \Delta T = W \cdot P$

This linearity is not obvious, because most of the heat is transferred by air convection (more than 90%), a process which in most cases is nonlinear. From eq.(5) it may be deduced that the cooling process following a short pulse with energy Q is exponential in time t:

(6) $\qquad \Delta T = Q/C_h \exp(-t/\tau)$

The time constant τ depends on the heat capacity C_h of the cone as follows:

(7) $\qquad \tau = C_h \cdot W$

According to the linear relation (4) and the linearity between ΔR and the output voltage of the Wheatstone bridge, we can also describe the output signal of the calorimeter by an exponential function:

(8) $\qquad U(t) = U_o \exp(-t/\tau) + U_{off} \quad$ with $U_o = a U_B Q/(4C_h)$

The additional term U_{off} stands for a possible offset of the bridge. Another time-dependent term describing a drift of the system can be omitted

as a result of the well-functioning compensation. The last equation suggests the following definition of the responsivity of the calorimeter with
reference to an electrical pulse of energy Q:

(9) $s_{el} = U_o/Q$

Finally, for the desired responsivity with reference to radiation pulses,
we have

(10) $s_{rad} = s_{el} \cdot \alpha \cdot F_1$,

where α is the absorptance and F_1 is the correction factor for the lead
heating. Another correction for a steady-state nonequivalence is neglected,
because both radiation and electrical heating happen within the volume enclosed by the bolometer winding.

This simple theory for the calorimeter does not take into account the finite thermal conductivity of the cone, the windings and the epoxy cement
used to bond the windings. Finite heat conductivity means that it needs a
finite time to balance the temperature inside the cone and its windings. It
is therefore only to be expected that eq.(8) describes the transient response for times larger than the balance time. Figure 4 shows the actual
calorimeter response to both radiation and electrical heating. (Please note
that three different time
scales are used to resol-
ve all details. The gaps
in the curves are inser-
ted to avoid the impres-
sion of breaks.) The dot-
ted line is the response
to radiation heating and
the solid line to elec-
trical   heating.   These
curves are free of noise
because they are averaged
over many individual re-
sponse curves. There is
obviously a considerable
disagreement   between
electrical and radiation
heating for times shorter
than 10 s. This is due to
a dynamic nonequivalence
which results from dif-

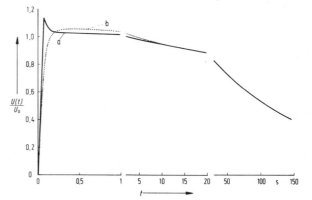

Fig. 4. Response   curves of the calorimeter
after (a) electrical   and   (b) laser heating
normalized to U_o. For the x axis three different time scales are used.

ferent transient heat currents as mentioned above. But both response curves
agree very well for times longer than about 15 s although they are independently normalized, which means that the dynamic nonequivalence has disap-
peared. The second pleasant feature is that the curves proved to be exactly
exponential. These two results justify the calibration prodecure described
above based on eqs. (8),(9) and (10). The evaluation of the measurements
can therefore be carried out in the following way: A least-squares fit to
eq. 8 is performed including all measurement points spaced more than 15 s
from the energy pulse. The fitting parameters are U_o, τ and U_{off}, although
only U_o is used to calculate the responsivity according to eqs. 9 and 10.

5. LASER ENERGY SCALE DERIVED FROM A RADIOMETER

The previous brief discussion has demonstrated how complicated a cone calorimeter is in both its thermodynamics and its handling; an easier method is
needed. Figure 5 shows such a method. The set-up is quite similar to that

used for the calibration of power meters but there is a fast electronic shutter inserted in the optical path. In this way is possible to cut off a pulse with a length of a few ms to several seconds from the cw laser beam. If the shutter is placed near the focus of the lens, the rise time of the pulse can be kept well below 1 ms. The reference detector, which is irradiated by the beam reflected on the front surface of the beam splitter, is calibrated at the beginning

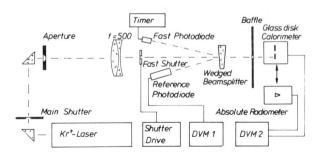

Fig. 5. Set-up for the transfer of the scale of the laser radiometer to a glass disk calorimeter.

by comparison with an absolute laser radiometer. In this step the shutter is continuously open. In the second step the radiant power is measured with the reference detector while the calorimeter is charged with a radiation pulse. The pulse duration can be measured with a fast photodiode receiving the rear surface reflection of the beam splitter and an electronic timer/counter with a relative uncertainty below 10^{-4}. The radiant power can be determined with an uncertainty not much larger than the scale uncertainty of the absolute radiometer, therefore the uncertainty of the energy scale should not be much larger than that for radiant power. The increase is mainly due to detector noise. It is obvious that the method can be reversed to derive a laser power scale from an absolute calorimeter. In both cases a serious difference between the shutter- produced pulses and the real laser pulses must be considered: The latter are much shorter. Investigations had therefore to be carried out to ascertain, whether the responsivity of glass disk calorimeters (Edwards 1970, Gunn 1973 and 1974, Xu 1979), which should be taken as secondary standards, depends on the pulse length. A favourable feature of these calorimeters is the high damage threshold with respect to irradiance. The reason for this is that the energy is absorbed in a volume instead of a surface, thus avoiding prohibitively high temperatures. In this study the ratio of the maximum output voltage U_{max} following a laser shot and the energy Q of the pulse was taken as the responsivity of the calorimeter:

(11) $s = U_{max}/Q$

It must be emphasized that the maximum output is a much more suitable output quantity than a voltage taken at a fixed time. The point of time for which the maximum occurs can shift considerably with increasing pulse lengths. Figure 6 shows that there is no significant change is responsivity in a range of pulse lengths from 10 ms to 500 ms. There

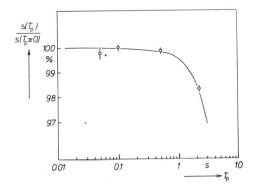

Fig. 6. Dependence of the responsivity s of the glass disk calorimeter from the pulse length. Circles are measurements, the solid line is calculated by convoluting the response curve for a very short pulse with pulses of duration T_p.

is no reason to doubt this constancy also for pulse lengths below 10 ms, which is the range of ordinary pulse lasers. This assumption is supported by the solid line in Fig. 6 which was computed by convoluting a response curve for a short pulse with the actual pulse length. It follows that commercial glass disk calorimeters can be used as secondary standards, but one must be aware of their limitations:

* The responsivity is not absolutely stable. Due to the blooming of the glass absorber the reflection losses, which are in the order of 4 %, may change (Blevin 1959).
* It is not possible to calculate the spectral responsivity from the spectral absorptance and reflectance, because the front surface heat loss depends on where within the volume the energy is absorbed. For example, the uv responsivity is strongly reduced because most of the radiation is absorbed just behind the front surface, which leads to increased convection losses.
* The responsivity is slightly nonlinear due to nonlinear convection losses.
* Electrical heating (if a heater is incorporated) and radiation heating are not sufficiently equivalent. The reason is, again, the poor heat conductivity of the glass disk. The heater should therefore only be used to check the stability of the temperature sensor.

6. INTERNATIONAL COMPARISONS

Up to now, two special international comparisons of laser power scales have been performed. The first was for He-Ne lasers at a power level of 3 mW (Honda, Endo 1978), the second for Ar^+ lasers at a power level of 200 mW (Endo, Honda 1983). Apart from a problem with the stability of the transfer standards the result was an agreement of the scales to within 0.5 %. Further comparisons of, for example, for CO_2 laser and for pulsed lasers are under consideration.

7. LITERATURE

Blevin W R 1959 Opt. Act. **6** 99
Boivin L P 1981 Metrologia **17** 19
Boivin L P 1982 Appl. Opt. **21** 918
Brandt F, Möstl K 1980 PTB Annual Report p. 157
Brusa R W 1983 Dissertation, Swiss Fed. Inst. of Techn. Zurich
Endo M, Honda T 1983 IEEE Trans. Instrum. Meas. **IM-32** 77
Edwards J G 1969 Opto-Electronics 1 452
Edwards J G 1970 J. Phys. E: Sci. Instrum. 3 452
Franzen D L, Schmidt L B 1976 Appl. Opt. **15** 3115
Gunn S R 1974 Rev Sci. Instrum. **45** 936
Honda T, Endo M 1978 IEEE J. Quantum Electr. **QE-14** 213
Möstl K 1978 Feinwerktechnik & Meßtechnik **76** 72
Möstl K 1988 Measurement (in press)
Möstl K, Brandt F, Xie X 1984 Proc. 11th Int. IMEKO Sym. Photon Detectors Weimar, p. 226
Xu D 1979 Report National Institute of Metrology China

Inst. Phys. Conf. Ser. No. 92
Paper presented at Int. Conf. Optical Radiometry, NPL, London, 12–13 April 1988

Factors limiting the accuracy of absolute radiometry

D. CROMMELYNCK
Institut Royal Météorologique de Belgique,
3, avenue Circulaire, B - 1180 Bruxelles

ABSTRACT: The general principles of absolute radiometry
are recalled in general. The magnitude of the uncertainties of the different factors or terms needed in the
radiometric equations are identified, in particular the
relatively poor accuracy of area measurement. Determination of the incertitude on parasitic effects and non-
equivalence is difficult to access. Comparisons of
independently designed and operated absolute radiometers
and manufacturing reproducibility remain good absolute
radiometry improvement indicators. The success of the
last Stefan Boltzmann determination demonstrates that an
accuracy of $1,5 \ 10^{-4}$ is feasible.

1. INTRODUCTION

Absolute radiometry is that part of metrology which is particularly difficult because its objective is to provide an
energy measurement information on sources of radiation which
depend on up to four variables. They are the spatial, spectral, directional and temporal distribution of the target
source, seen eventually through an absorbing medium. The
resulting difficulty to compare two sources, causes sensor
calibration methods with reference sources to be not really
appropriate and misleading.

This led to the concept that radiation metrology should be
based on absolute radiometric sensors. Their design aims at
providing well defined measurements based on instrumental
radiometric equations. The absolute radiometers can be operated in different conditions, air or vacuum, low temperature
or laboratory temperature and will behave accordingly with
more or less well known accuracy ot the terms and factors of
their radiometric equation.

The accuracy which can be obtained with absolute radiometric
measurements depends not only on the radiometer design but
mainly on the nature of the source. The integration of the
spectral distribution of the radiative energy reduces the
difficulty to the accurate knowledge of the blackness of the
sensor. Sufficiently stable irradiances from uniform or
punctual sources like the Sun can be measured with a

relative accuracy, which is still improving, when the aperture angle is small enough.

If the radiometer can be designed to have a relatively fast rise time of up to 1 sec. at 10^{-3} level, it is excluded to make corrections when the source is not uniform and when the departure from the cosine response is unacceptable.

2.   THE BASIC PRINCIPLE OF ABSOLUTE RADIOMETRY

Absolute radiometry is based on the simultaneous or successive comparison of two identical thermal fields excited respectively by the unknown radiation energy and the measurable electrical energy. The unavoidable difference in nature of the two energy sources leads automatically to the comparison of two not strictly equivalent thermal fields due to the difficulty to transform completely radiation into heat and the difference in heat distribution. The comparison of the two thermal fields can be done either by temperature or by differential heat flux measurements with respect to a common thermal reference.

Bridge like thermal fields methodology should be favoured because it provides a linear behaviour, a symmetry in space and time, a protection against identical perturbations and test possibilities with different measurement configurations. The balancing can be done manually for steady state excitation of the thermal field or automatically, together with the usage of appropriate data acquisition techniques.

Accessories are needed to take care of the specificity of the radiation source,its spatial distribution as well as its nature, be it radiance or irradiance. In this last case a strictly uniform flux distribution is required at the sensitive aperture.

Being absolute the radiometer is described by a set of equations whose coefficients and factors need to be determined with the highest accuracy possible. As their values could eventually change with time it is necessary to provide wherever possible a radiometric design allowing their direct laboratory determination. This is called radiometric characterization. Implicit assumptions should be avoided.

3.   IDENTIFICATION OF THE LIMITATIONS OF ABSOLUTE RADIOMETRY

The realisation of the principle of absolute radiometric measurements can be based on different designs, all of them require a series of determinations which are accuracy limited. Let us consider them.

3.1 Thermal fields zero or balance detection

The accuracy of a radiometric measurement cannot be better than the thermal field unbalance incertitude. Each design should be optimized with that objective serving as a grading index. With semiconductor heat flux meters, sensitivities of 125 $\mu V/mW$ can be achieved which enables in air resolutions of up to 5.10^{-5} for an input level of 140 mW at 293 K temperature. In vacuum and at a temperature of 5K a resolution of better than 10^{-6} can be achieved at a 1 mW level with a germanium resistance thermometer. Vacuum and cryogenic temperature operation results in more than two orders of magnitude improvement in sensitivity.

3.2 Electrical heating accuracy

The electrical measurable power to which the absorbed radiative power is compared is given by V_h . V_r/R where V_h is the voltage over the heater and V_r the voltage over the reference resistor R. The relative accuracy on the voltage measurement at the range of 3V is about +/- 1,5 10^{-5} and +/- 5.10^{-5} for the resistance at the level of 100Ω. The resulting power accuracy will thus be known with an accuracy of +/- 8.10^{-5}. When the unknown radiative power varies, V_h and V_r/R should of course be measured simultaneously with the zero balance maintained.

3.3 Determination of the radiation converted to heat

To absorb the radiation on a black painted surface is usually not enough because the accuracy of the absorption factor is only of the order of +/- 10^{-2}. To improve the situation benefit is taken from the geometrical enhancement obtained with cavities at the expense of loosing equivalence on the spatial distribution, of the heated area .

The cavity can take a variety of shapes depending on the nature of paint, diffuse or specular. It can be provided with mirror surfaces giving optical feedback. Heaters are generally localised in front of the precise sensor aperture. Very good results are obtained with specular black surfaces giving effective cavity absorptions as high as 0,99998. With diffusing 3M Velvet black the effective absorption is less high for comparable geometries, but it can be measured with an incertitude less than 10^{-4}.

The fact that the cavity absorbs all the radiation does not mean that it is totally measured (leakage) and exactly compared to the electrical energy. It is thus necessary to introduce the concept of cavity or sensor "efficiency" which establishes in function of the conditions (vacuum or air operation, temperature) the ratio of radiation converted to heat that is really measured and compared to the electrical power as it is distributed within the cavity, in relation with the zero or balance detection. The configuration of the cavity with its heater and balance detector (heat field) can be very different from one design to the other ; further developments are needed to qualify the corrections and the determination of the non equivalence factors in the conditions of use.

Especially for this it is important to organize radiometric comparisons on the Sun or other common stable sources (a laser beam can be stabilized to 5.10^{-5}/h) with instruments developped really independently.

3.4   Incertitude in the sensitive area

For irradiance measurements of an uniformly distributed flux the area of the sensor sensitive aperture needs to be known with a relative error that should not be higher than those with the electrical measurements.

Is this trivial or not so easy to achieve ? To answer this question two stainlessteel holes were circulated among eight metrology laboratories which kindly accepted to measure their diameter in two orthogonal directions. The results summarised in table 1 shows relative differences as high as 10^{-3} for an hole of 5 mm diameter. This is five times higher than the best incertainty claimed.

The resulting uncertainty on the area at the 99% level approaches +/- 10^{-3}. As the absolute incertitude remains about the same for larger apertures it is thus necessary to increase the diameter to improve the relative accuracy of irradiance measurements.

The manufacturing of the apertures is a critical item in particular with respect to the circularity and the edges which should be sharp and cyclindrical.

3.5   Miscellaneous parasitic effects

According to the type of absolute radiometer different kind of corrections need to be performed and effects taken into account.
-   A temperature correction on the area of the sensitive aperture depending on the material used.
    Manipulation and long term changes were reported.
-   Reflections at the aperture lands.
-   Pointing error.
-   Heating effects of the sensitive aperture which can be reduced with a high reflectance spherical shaped mirror aperture and a small slope angle ; air operation requires a higher correction than for vacuum.
-   Scattering of thermal and optical radiation from the radiation trap and front aperture depending on the optical configuration of the sensitive aperture.
-   Diffraction losses.
-   Losses and/or dissymetry of electrical power leads heating
-   Servo system error.
-   Response time of the radiometric system in the case of varying source.
-   Main aperture shutter emission correction.

Table 1.   COMPARISON OF DIAMETER MEASUREMENTS OF TWO REFERENCE HOLES OF NOMINAL DIAMETER 5 AND 8.5 mm

Metrological service (Instrumentation)	Incertitude +/-	Temperature °C	Angle from origin 0°	Angle from origin 90°	Angle from origin 0°	Angle from origin 90°
1. Service de la Métrologie (Ministère des Affaires Economiques (B)) (Mesure interférométrique laser + microscope universel)	0.002 (95%)	20	5.018	5.018	8.504	8.503
2. Test Services Division (ESTEC/ESA (EUR)) (Laser interferometer + SIP "Trioptic")	0.001	20.2	5.0229	5.0237	8.5060	8.5058
3. Technische adviseurs B.V. (NL) (Universele 3D meetmachine ZEISS UMM500)	0.001	20	5.021	5.023	8.506	8.507
4. Bundesamt fur messewesen-Berne (CH)	0.001	--	5.0194	5.0193	8.5045	8.5048
5. National Physical Laboratory (G B)	0.0005	20	5.0210	5.0212	8.5058	8.5055
6. Council for Scientific and Industrial Research (South Africa) (SIP MU214B + Tesa lever)	0.001 (99,7%)	20	5.0207	5.0209	8.5059	8.5049
7. Division of Physics National Research Council (CND) (Comparateur optique - Topic 14, Jones and Lamson)	0.0018 0.0015	23	5.0178	5.0195	8.5011	8.5001
8. National Bureau of Standards Department of Commerce (US)	0.0005 (3σ)	20	5.0205	5.0210	8.5056	8.5054
Mean diameter			5.0202	5.0208	8.5049	8.5046
+/- uncertainty (99%)			0.0021	0.0024	0.0021	0.0026

4.   CONCLUSIONS AND STATE OF THE ART

Important improvements have been performed in absolute radiometry during the last decade.

Work at low radiation levels and laboratory standards maintenance are best performed with cryogenic absolute radiometers.   However the basic incertitudes on area and electrical heating are the same for the now classical cavity absolute radiometers working in air or vacuum and the cryogenic radiometers.   Larger apertures will not only allow to reduce the area incertitude but will enable easier laboratory characterization.

Althought vacuum operation allows to reduce corrections due to thermal leakage, it is very useful to know the air to vacuum sensitivity ratio because this enables to crosscheck measurements for consistency.   If the relative accuracy on the Solar irradiance is actually not better than +/- 0,15-%, the recent radiometric determination of the Stefan - Boltzmann constant demonstrates that an improvement of one order of magnitude is feasible.   Further research and improvements are necessary not only for small aperture angles but even more for absolute radiometry with larger aperture angles on extended and eventually varying sources.

5.   REFERENCES

Brusa R W and Fröhlich C 1986 Appl. Opt. 25, 4173-4179
Crommelynck D 1973   IRM. Publ. Série A 81, 1-50
Crommelynck D 1982 NASA. Conf. Publ. 2239
   (Virginia : LARC), 53-71
Geist J 1972 NBS Techn. Note 594-1
   (Washington DC : NBS), 1-54
Gillham E J 1962 Proc. R. Soc. A 269, 249-276
Kendall J M and Berdahl C M 1970 Appl. Opt. 9, 1082-1091
Martin J E, Fox N P and Key P J 1985 Metrologia 21, 147-155
Quinn T J and Martin J E 1985 Phil. Trans. R. Soc. Lond.
   A 316, 85-189
Willson R C 1973   Appl. Opt. 12, 810-817.

Inst. Phys. Conf. Ser. No. 92
Paper presented at Int. Conf. Optical Radiometry, NPL, London, 12–13 April 1988

The electron storage ring BESSY as a primary radiometric standard

M. Kühne

Physikalisch-Technische Bundesanstalt, Abbestr. 2-12, D-1000 Berlin 10, Germany

ABSTRACT: The electron storage ring BESSY was investigated in respect of its use as a primary radiometric standard. From the knowledge of the storage ring parameters electron energy, magnetic induction, electron current, and with proper corrections for the electron beam size and electron beam divergence the spectral flux through an aperture can be calculated with an uncertainty of 0.23% in the visible and near infrared and 2% in the soft X-ray region at 5 keV photon energy.

1. INTRODUCTION

The availability of radiometric primary source standards is of fundamental importance for the field of radiometry. Primary source standards by definition are sources of which the radiation emission can be calculated from the principles of physics with the knowledge of a few basic parameters. The blackbody radiator is an example for a primary standard as its spectral emission can be calculated using Plancks law and is dependent only on the blackbody temperature.
For technical reasons the maximum achievable temperature of such a blackbody radiator is limited to about 3000 K. Consequently the short wavelength limit for radiometric application of blackbody sources is in the UV at about 250 nm.

For shorter wavelengths for a long time no primary sources with sufficiently small uncertainties were available. Wall stabilized noble gas arcs doped with other elements to produce optically thick lines which approach the Planck function according to the temperature of the arc have been used (Boldt 1961) as well as hydrogen arcs (Ott et al 1975) of which the continuum emission can be calculated from theory. All these sources have the disadvantage that the achievable uncertainty in the radiation emission is insufficient and that they can cover only a limited spectral range. Below 105 nm where no solid window material is available only large differential pumping systems allow to extend the operation to about 53 nm (Behringer and Thoma 1979) and for even shorter wavelength no primary standards were available at all.

2. THE ELECTRON STORAGE RING BESSY AS A SOURCE OF CALCULABLE SPECTRAL RADIANT POWER

The development of electron synchrotrons in the mid forties led to the theory of the radiation emission of relativistic electrons moving through a homogenous magnetic field (Schwinger 1949). As synchrotron radiation sources are emitting under vacuum conditions they are well suited for VUV and soft X-ray radiation applications. A typical synchrotron radiation spectrum (Fig. 1) extends from the IR through the VIS and UV to the VUV and soft X-ray region. Electron synchrotrons have been used for radiometric work, but not as primary standards.

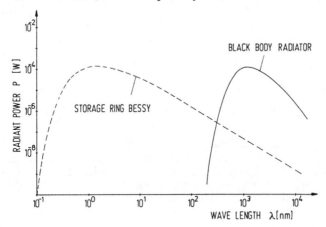

Fig. 1: Comparison of the radiant power of BESSY and a blackbody radiator. BESSY: Energy W=800 MeV, current I=20 mA, bending radius ρ=1.79 m, $\Delta\lambda/\lambda=10^{-3}$, aperture stop 40 x 40 mm^2 at 7.5 m. Blackbody radiator: T=3000 K, circular emitting area with a diameter of 10 mm, $\Delta\lambda/\lambda = 10^{-3}$, aperture stop 40 x 40 mm^2 at 1.0 m.

With the availability of electron storage rings work began at several places like NBS in the USA (Madden et al 1985), ETL in Japan (Suzuki 1984), the Institute of Nuclear Physics in the Soviet Union (Gluskin et al 1980), and at PTB in Germany (Kühne et al 1983), to investigate their use as primary radiometric standards. In an electron storage ring the fundamental parameters required for the spectral radiant power calculation, the electron energy W, the magnetic inductance B in the bending magnet at the point of observation and the electron current I stored in the ring can be precisely determined. The spectral radiant power $\Phi_\lambda(\lambda)$ through an aperture A in a distance d^{SR} at an angle Ψ_0 to the electron orbit plane (see Fig. 2) is given by:

$$\Phi_\lambda^{SR}(\lambda) \;=\; \Phi_\lambda^{SR,\parallel}(\lambda) + \Phi_\lambda^{SR,\perp}(\lambda) \;=$$

$$\frac{2e\rho^2 I}{3\varepsilon_0\lambda^4\gamma^4 a d^{SR}}\left\{\int\limits_{\psi_0-\psi'}^{\psi_0+\psi'}[1+(\gamma\psi)^2]^2\,K_{2/3}^2(\xi)\,\mathrm{d}\psi + \int\limits_{\psi_0-\psi'}^{\psi_0+\psi'}[1+(\gamma\psi)^2]\,(\gamma\psi)^2\,K_{1/3}^2(\xi)\,\mathrm{d}\psi\right\}$$

with

$$\gamma = \frac{W}{m_0\,c^2}\,, \qquad \xi = \frac{2\pi\rho}{3\gamma^3\lambda}[1+(\gamma\psi)^2]^{3/2}\,, \qquad (\,1\,)$$

$$\rho = \frac{W}{ecB}\,, \qquad \psi' = \frac{a}{2d^{SR}}\,.$$

At the electron storage ring BESSY the electron energy W is determined by the spin depolarisation method (Lehr and Isoyama 1985). The magnetic inductance B is measured with an NMR-probe. The electron current I is determined in the range from 1 mA to 1000 mA using two identical toroidal dc beam current transformers ,see (Riehle and Wende 1986) for further details. Very small currents consisting of 1 electron up to about 1000 electrons (1 electron at BESSY constitutes a current of about 0.77 pA) can be determined by electron counting (Riehle et al 1988). For the intermediate range from about 1 nA to 1 mA an optical system consisting of several solid state diodes is used (Bernstorff et al 1987).

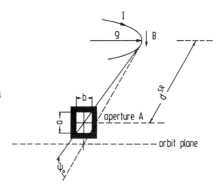

Fig. 2: For radiometric measurements the flux Φ is limited by the aperture stop A (area a x b) in the distance d^{SR} from the tangent point. Ψ_o is the angle enclosed between the electron orbit plane and the center of the aperture stop.

Equation (1) does not consider the vertical extension of the stored electron current or its angular divergence as it assumes that all electrons travel on the same ideal circular orbit. Especially in the soft X-ray region this approximation is not valid any more and corrections to equation (1) may be required, depending on the size and location of the flux limiting aperture, see (Riehle and Wende 1987).

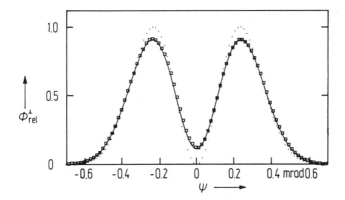

Fig. 3: Angular distribution of the vertically polarized component of the synchrotron radiation of BESSY at photon energy of 2.78 keV. The electron energy is 755 MeV, the stored current I = 130 mA. The squares represent the measured distribution, the dashed curve is the calculation according to Schwinger with infinite small beam extension and beam divergence. The solid line is the calculation with the beam size and beam divergence considered.

The vertical electron beam size is measured using a x-ray slit camera working at photon energy of 2.78 keV. The vertical electron beam divergence is determined with a bragg polarimeter that scans the angular distribution of the synchrotron radiation. A description of the system is found in Riehle (1986).The angular distribution of the vertically polarized component of the synchrotron radiation emitted by BESSY at a photon energy of 2.78 keV is shown in Fig. 3.

The uncertainties in parameters required for the determination of the spectral photon flux of BESSY and the resulting uncertainties are listed in table 1.

Table 1:

Uncertainties $\Delta\Phi/\Phi$ of the spectral photon flux of the storage ring BESSY ($\sqrt{3}\ \sigma$ level) at the photon energies 1 eV and 5 keV.

Source of uncertainty	relative uncertainty $\Delta\Phi/\Phi$	
	1eV	5keV
magnetic induction B = (1.4077 +- 0.002) T (measured by an NMR probe)	0.1%	1.2%
electron energy W = (754.5 +- 0.2) MeV	0.02%	0.55%
electron current I = (10 +- 0.005) mA	0.05%	0.05%
distance d^{SR} = (6400 +- 5) mm	0.16%	0.16%
aperture A = (4 +- 0.004) mm² (E = 1 eV) area A = (4 +- 0.016) mm² (E = 5 keV)	0.1%	0.4%
emission angle Ψ_o = (0 +- 8) μrad	0.002%	0.14%
vertical extension FWHM = (390 +- 60) μm and divergence FWHM = (100 +- 30) μrad	0.002%	1.5%
sum in quadrature $\Delta\Phi/\Phi$ (total)	0.23%	2.0%

3. EXPERIMENTAL VERIFICATION

In a first experiment a comparison of the calculated spectral irradiance of BESSY was performed against the spectral irradiance of a tungsten halide lamp that was calibrated using a blackbody radiator. The comparison showed good agreement (Riehle and Wende 1986) but due to the uncertainty in the halide lamp calibration it was not possible to confirm the small uncertainty of less than 1% for the synchrotron radiation emission.

To utilize the small uncertainty in the spectral responsivity scale obtained by the NPL cryogenic radiometer, it was decided to intercompare the radiometric scale based on the cryogenic radiometer at NPL and the scale given by the electron storage ring BESSY. For that purpose a transportable transfer radiometer was built and a joint NPL/PTB experiment performed.

The transfer radiometer consisted of a filter-detector system (see Fig. 4) with a Glan-Thompsen calcite polarizer (extinction ratio of 1 in 10^6) with a silicon diode as detector with an area of 300 mm^2. Two different interference filters with peak transmissions at 676 nm (bandpath 24 nm FWHM) and 802 nm (bandpath 18 nm FWHM) were used .The calibrations were performed at 676.4 nm and 799.3 nm to utilize a krypton-ion laser at NPL.

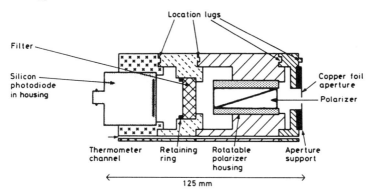

Fig. 4: Schematic of the transfer radiometer

At the radiometric laboratory of PTB the transfer radiometer was calibrated using BESSY as a primary standard and the two spectral responsivity scales compared. The uncertainties in the calibation of the spectral responsivity scales are shown in Table 2:

Table 2:

Uncertainty of the intercomparison of the radiometric scale based on the NPL cryogenic radiometer and on the electron storage ring BESSY.

Uncertainty of absolute spectral responsivity $S^{RAD}(\lambda)$ determined via the cryogenic radiometer	0.15%
Uncertainty of absolute spectral responsivity $S^{SYN}(\lambda)$ determined via BESSY	0.35%
Combined uncertainty	0.38%

For the calibration of the transfer radiometer the storage ring was operated with stored electron currents between 9 mA and 300 mA. Both in the VIS and the near IR the mean value of the calibration at BESSY differed from the calibration value obtained at NPL by only 0.1%. The standard deviation (1σ value) for both wavelength was also 0.1%.

For the verification of the calculated uncertainty of the spectral emission of BESSY in the soft X-ray region in a first experiment a comparison with a radioactive ^{57}Co standard was performed. The known radiation emission of the ^{57}Co standard (uncertainty 1%) at 6.4 keV was used to calibrate the absolute spectral responsivity of a Si(Li)-detector in the soft X-ray region. The spectral radation emission of

BESSY at 3 keV was then determined using the calibrated Si(Li)-detector. The uncertainty achieved in this preliminary calibration was about 1%. At present a more accurate experiment is under way based on the emission of a ^{51}Cr standard.

4. CONCLUSION

The electron storage ring BESSY can be used as a primary radiometric standard from the near IR down to the soft X-ray region. The uncertainty for the calculation of the spectral radiant power through an aperture is 0.23 % in the near IR and the VIS and increases to about 2% in the soft X-ray region. An intercomparison with the radiometric scale based on the NPL cryogenic radiometer and BESSY brought good agreement with in the combined uncertainty of 0.38% of the intercomparison.

REFERENCES

Behringer K and Thoma P 1979 Appl. Opt. **18** 2586
Bernstorff S, Hänsel-Ziegler W, Ulm G, and Wolf F P 1987 BESSY Annual Report 99
Boldt G 1961 Proc. 5th Int. Con. Ion. Phen. in Gases. H.Maecker (ed.) Vol.I, 925-939. Amsterdam; North Holland
Fox N P, Key P J, Riehle F, and Wende B 1986 Appl. Opt. **25** 2409
Gluskin E S, Trakhtenberg E N, Feldman I G, Kochubei V A 1980 Space Sci. Instrum. **5** 129
Ott W R, Behringer K, Gieres G 1975 Appl. Opt.**18** 2121
Kühne M, Riehle F, Tegeler E, and Wende B 1983 Nucl. Instr. and Meth. **208** 399
Lehr H and Isoyama G 1984 BESSY Annual Report 32; see also Serednyakov S I, Skrinskii A N, Tumaikin G M, and Shatunov Yu M 1976 Sov. Phys. JETP **44** 1063
Madden R P, Ederer D L, and Parr A C 1985 Nucl. Instr. and Meth. **B10/11** 289
Riehle F 1986 Nucl. Instr. and Meth. **A246** 385
Riehle F and Wende B 1986 Metrologia **22** 75
Riehle F and Wende B 1987 Optik **75** 142
Riehle F, Bernstorff S, Fröhling R, and Wolf F P 1988 Nucl. Instr. and Meth. in press
Schwinger J 1949 Phys. Rev. **75** 1912
Suzuki I H, 1984 Nucl. Instr. and Meth. **228** 201

Inst. Phys. Conf. Ser. No. 92
Paper presented at Int. Conf. Optical Radiometry, NPL, London, 12–13 April 1988

31

The intercomparison of two cryogenic radiometers

N.P.Fox and J.E.Martin

Division of Quantum Metrology, NPL, Teddington, UK.

ABSTRACT: Two cryogenic radiometers (electrical substitution radiometers cooled to liquid helium temperatures) have been developed at NPL. The first was used to realise thermodynamic temperature and to determine the Stefan-Boltzmann constant. The latter measurement confirmed the calculation that the radiometer is an absolute instrument with an uncertainty of 0.02%. The second radiometer was dedicated to optical radiation measurements with a calculated uncertainty of 0.005%. This paper describes and presents results of the intercomparison of the two radiometers by comparing the responsivity of silicon photodiodes as independently determined by the radiometers.

1. INTRODUCTION

A cryogenic radiometer, that is, an electrical substitution radiometer cooled to liquid helium temperatures, has been designed and constructed at NPL to measure the total radiative power from a black body in the temperature range -130 to +100°C (Quinn and Martin 1985, Martin *et al* 1988). The calculated uncertainty in the measurement of this power is about 100 parts per million, and reduces to between 2 and 3 parts per million when measuring the power ratio from the black body at two different temperatures. The work had two objectives, firstly, to measure thermodynamic temperature and secondly, to make an absolute determination of the Stefan-Boltzmann constant, σ. Both these objectives were achieved but it is only the latter that is of direct significance to this present work.

The value of σ, as measured by Quinn and Martin (1985) is
$$5.66959 \pm 76 \times 10^{-8} \text{ W m}^{-2} \text{ K}^{-4}$$
compared to the recommended value of
$$5.67051 \pm 19 \times 10^{-8} \text{ W m}^{-2} \text{ K}^{-4}$$
as derived by the Codata Task Group on Fundamental Constants (1986) from the equation
$$\sigma = 2\pi^5 k^4 / 15h^3 c^2$$
where k is the Boltzmann constant, h is the Planck constant and c is the speed of light. The difference between the two values is 1.6 parts in 10^4, this is slightly more than their combined standard deviation (1.4 parts in 10^4) and therefore they can be considered not to be in disagreement. Thus, this experimental determination of σ confirms the calculations that the radiometer is an absolute radiometric detector with a maximum uncertainty of 1.6 parts in 10^4. The uncertainties quoted throughout this paper are at the one standard deviation level.

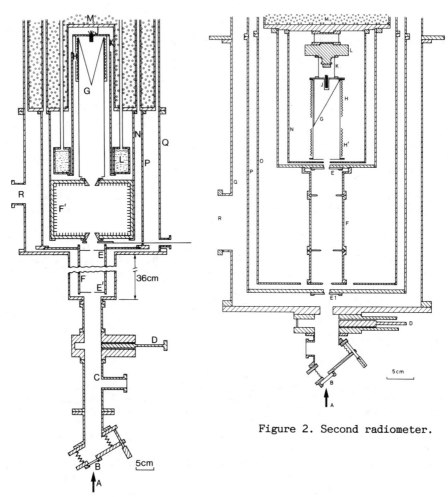

Figure 2. Second radiometer.

Figure 1. Original radiometer.
The legend applies to both figures. A - laser beam, B - Brewster-angle
window, C - lower vacuum chamber, D - gate valve, E and E' - scatter
detectors, F and F' - radiation traps, G - cavity, H and H' - heater, K -
heat link, J - germanium resistance thermometer, L - constant temperature
heat sink (figure 1, a superfluid helium bath, figure 2, a copper block),
M - helium reservoir, N,O and P - copper tubes/radiation shields, Q -
outer vacuum chamber, R - pumping port.

Following the success of this radiometer a second cryogenic radiometer,
dedicated to the measurement of optical radiation, was designed by Martín
et al (1985) and constructed by Oxford Instrument Ltd. The principal
function of this radiometer is to provide an absolute radiometric scale
against which secondary transfer standards can be calibrated, for example,
silicon photodiodes. To realise the scale the radiometer is used to
measure the power of an intensity-stabilised laser with a calculated
uncertainty of 5 parts in 10^5. Obviously this uncertainty will be

substantiated if it can be confirmed experimentally by comparing the power measurement with a similar measurement using the original radiometer.

This paper describes the comparison, that is, the measurement of the power of an intensity-stabilised helium-neon laser, in the range 0.6 to 1.0 mW, by the two radiometers with an uncertainty of between 1 and 2 parts in 10^4.

2. THE RADIOMETERS

The design and construction of the radiometers, including the operating principles, have been fully described in the earlier papers.

Figures 1 and 2 show the line diagrams of the original radiometer with modifications and the second radiometer respectively.

For this comparison the original radiometer has been modified to measure the power of a laser beam instead of the power from a black body. To permit the laser beam to enter the radiometer cavity, the black-body, which was suspended inside the vacuum chamber, has been removed and the lower vacuum chamber replaced with a new vacuum chamber, C, incorporating a Brewster-angle window, B. The radiation shields, which originally surrounded the black body, have also been removed and replaced with one shield, F, to which the scatter detectors, E and E', are attached. This shield is made of copper and restricts thermal radiation from other sources entering the cavity. The shield is cooled to about 80 K by attaching it to a copper flange at the end of a tube, P, which is anchored to the liquid nitrogen reservoir.

The second radiometer was designed to measure laser power and therefore required no modification.

3. METHOD OF COMPARISON

The comparison can be undertaken in various ways. The ideal method is to measure the laser power directly by keeping the optical path of the beam fixed and aligning the radiometers in turn, so that the beam passes directly into the absorbing cavity. Unfortunately, the radiometers are not portable instruments and it is not practical to pursue this method.

An alternative method is to keep the radiometers in a fixed position and to divert the beam into each cavity in turn by adjusting a mirror. However, recent work at NPL has shown that the scatter of light from a mirror cannot be neglected and its measurement over a complete hemisphere can only be made with great difficulty. Also, it cannot be easily demonstrated that the reflectivity of the mirror would remain identical when its orientation is changed. These considerations led us to believe that the uncertainty in the comparison would be at least 5 parts in 10^4 and this method was also abandoned.

The method adopted was to compare the responsivity of a set of silicon photodiodes after they had been independently calibrated against the two radiometers. This method does rely on the short term stability of the diodes but has the distinct advantage that the laser beam can be aligned separately for the two comparisons. It was thought that this method would achieve the required uncertainty.

The photodiodes chosen as the transfer standards for this intercomparison were Hamamatsu types S1227-1010 and S1337-1010. These devices were chosen because previous investigations at NPL have shown them to be spatially uniform, linear and stable with time. The devices differ from each other in their responsivity characteristics, type S1337 is a red enhanced device and has a near unity internal quantum efficiency in the visible spectral region, whereas, type S1227 has been doped to reduce its responsivity in the near infrared and consequently throughout the rest of the spectrum.

Figure 3 shows the optical and electrical system for calibrating the silcon photodiodes using the radiometers. The identical optical system is used with both radiometers. However, the instrumentation used for the electrical measurements is different.

The procedure for the calibration is as follows. The window is first cleaned and its transmittance measured before it is bolted on to the vacuum chamber. The vacuum chamber is then evacuated until the pressure is low enough for the gate valve to be opened. The laser beam is directed via a limiting aperture of about 7 mm diameter into the cavity with a mirror. The beam is aligned to pass along the vertical axis of the cavity by adjusting the mirror until the signal on the scatter detectors is a minimum. The window angle is then adjusted to the Brewster angle. The absorption of the beam in the cavity causes its temperature to rise above that of the reference temperature heat sink, the temperature of the cavity being monitored with a germanium resistance thermometer. When thermal equilibrium has been achieved the laser beam is interrupted with a shutter. Electrical power is then substituted in the cavity and adjusted

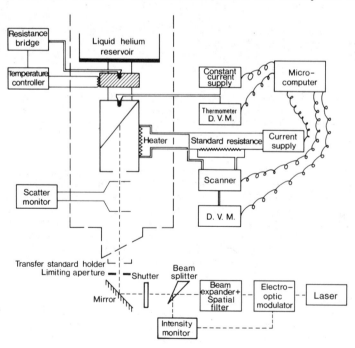

Figure 3. Block diagram of optical and electrical system for calibrating transfer standard detectors.

until the temperature rise of the cavity is identical to that when absorbing the laser beam. The electrical power is then equal to the power of the laser beam after correcting for the absorptance of the cavity, the transmittance of the window and any residual laser radiation detected on the scatter detectors.

Whilst the electrical power measurements are being made, the silicon photodiodes to be calibrated are positioned in the laser beam just below the window and above the limiting aperture. Their responsivity is then determined. The final part of the calibration procedure is to remove the window and re-determine its transmittance.

The following table gives the corrections and the uncertainties that are applied to the measurement of the laser beam power for the radiometers. Values are given as parts in 10^4.

	Original Radiometer		Second Radiometer	
	Correction	Uncertainty	Correction	Uncertainty
Window Transmittance	2.5	0.5	2.5	0.5
Beam Scatter	1.0	0.15	0.5	0.15
Cavity Absorptance	1.9	0.5	0.2	0.1
Electrical Power	-	0.1	-	0.1
Radiometer Sensitivity	-	0.1	-	0.3
Miscellaneous	-	0.1	-	0.1

4. RESULTS and DISCUSSION

The changes in the responsivity for two sets of three windowless silicon photodiodes, Hamamatsu types S1227 and S1337, are shown in figures 4a and 4b.

It can be seen that for the three photodiodes type S1227 the average change in the responsivity as determined by the two radiometers is 11 parts in 10^4, whereas, the average change is only 6 parts in 10^4 for measurements using type S1337. Various theories have been put forward to explain these results and investigations have been made into the performance of the diodes for changing atmospheric conditions, for example, the effect of humidity on the surface charge and their responsivity temperature coefficients. However, no solution can at present be offered as to why the two sets should differ. It should be stressed that the radiometers are not in the same room and that it is difficult to maintain equivalent environmental conditions, for example, it is not possible to completely 'black-out' the room containing the original radiometer.

One possible source of uncertainty arises in their surface reflectance. The reflectance of these diodes measured in a separate experiment is about 30% and assumed to be specular. The accepted practice when making reponsivity measurements is to align the diode with respect to the incident beam so that the path of the reflected beam is as close to the incident beam without interfering with the laser stabilisation optics. This optical arrangement prevents the reflectance of the diodes being measured simultaneously with the responsivity measurements and hence small changes in the reflectance could go undetected.

More recently further responsivity measurements using the original radiometer only have been made on the same diodes, a set of windowless E G

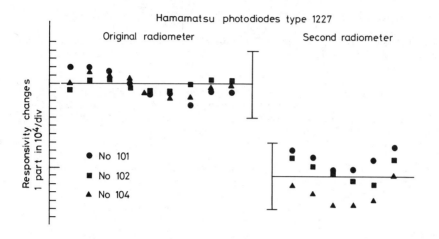

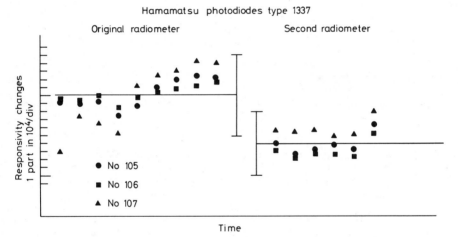

Figures 4a and b. Changes in the responsivity of two sets of Hamamatsu photodiodes as measured by the two radiometers. The measurements were made over a period of seven days with a two day interval between the measurements in the original and second radiometer. The error bar represents the combination at the one standard deviation level of the uncertainty in the laser power measurement (table 1), the random uncertainty of the photodiodes measurements and the systematic uncertainty (3 parts in 10^4) in using Hamamatsu photodiodes.

and G type UV444B with a surface reflectivity of about 10%, a wedged-window Centronic diode type 300-5 and a combination device of three Hamamatsu diodes type S1337. The latter device follows the design suggested by E.Zalewski at NBS to reduce the problems associated with the reflected beam by arranging the diodes so that the reflected beam only emerges from the device after five internal reflections and is therefore of reduced significance. The results of these measurements are shown in figure 5.

Again the diodes have behaved erratically although there appears to be an overall shape to the results. This shape maybe caused by a drift in the

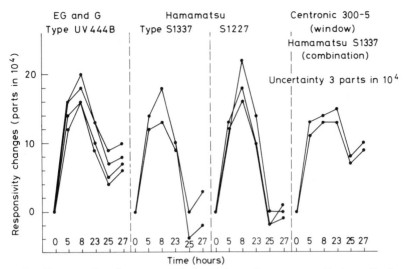

Figure 5. Changes in the responsivity of various sets of photodiodes.

electronics associated with the photodiodes. However, the relative change in response between the various sets would still suggest some undetected surface effect. Also, because of the smaller scatter associated with the Centronic and the combination device, it is possible that scatter radiation may also be affecting the diodes, that is, our present techniques for measuring the background radiation are not adequate. Work is continuing to try to solve these problems.

5. CONCLUSIONS

The radiometers are in agreement within the uncertainty of the comparison, that is, 4-5 parts in 10^4. This uncertainty stems directly from the photodiodes and it seems unlikely that this uncertainty can be improved to better than 2 parts in 10^4. Hence a more direct method of comparison is required. This may well be possible with the next generation of cryogenic radiometers if they can be made more mobile.

6. REFERENCES

Codata Bulletin 1986 **63** pp11-12
Martin J E, Fox N P and Key P J 1985 *Metrologia* **21** pp 147-155
Martin J E, Quinn T J and Chu B 1988 *Metrologia* **25** pp 107-112
Quinn T J and Martin J E 1985 *Philos. Trans. R. Soc. London* (Ser A) **316** pp 85-189

Post-conference note. It has since been discovered that the helium-neon laser used for the intercomparison has been emitting laser radiation at a number of wavelengths other than 632.8 nm and at a power level of a few tenths of a micro-watt. This extraneous radiation, which is not necessarily intensity-stabilised, may well be seen differently by the two radiometers as well as the photodiodes which have different responsivity characteristics and so could account for some of the effects described in this paper.

Inst. Phys. Conf. Ser. No. 92
Paper presented at Int. Conf. Optical Radiometry, NPL, London, 12–13 April 1988

Superconducting inductance bolometer with potential photon-counting sensitivity: a progress report*

J.E. Sauvageau and D.G. McDonald

National Bureau of Standards, Boulder, Colorado 80303

ABSTRACT: This bolometer is based on the temperature dependence of the inductance of a superconducting microstrip line. Since the device is superconducting it has no Johnson noise. It can be impedance matched to an optimized SQUID preamplifier, the quietest of all amplifiers, and its bias current is relatively unrestricted by self heating. We show theoretically that this device can have a sensitivity comparable to that of an optical photon counting detector, or an NEP_e of 5.4×10^{-18} W/$\sqrt{Hz}$. Our experimental prototype device is designed to test the theory of operation, but not at the highest levels of sensitivity.

1. INTRODUCTION

Conventional bolometers are limited in performance by their intrinsic Johnson noise, phonon noise, amplifier noise and self heating. Recently, a novel superconducting thermometer for bolometric applications was proposed by McDonald (1987). A theoretical analysis of this device suggested that a bolometer incorporating this temperature transducer can be very sensitive, an important consideration for infrared thermal detectors. To emphasize this sensitivity we will argue that this device can be made as sensitive as optical photon counting detectors. It is not our intention to suggest such an application, but rather to suggest that thermal detectors can potentially be as sensitive as any form of detector, contrary to the popular view.

We will motivate our study with a brief review of some high quality bolometers and then discuss some concepts, the superconducting penetration depth and kinetic inductance, needed to explain our device design. We then discuss our experimental work on a kinetic inductance thermometer, which is to be incorporated into a bolometer, and conclude with two comparisons of our bolometer with other devices.

2. RESISTIVE BOLOMETERS: A Brief Review

As depicted in Fig. 1, a bolometer is a device which typically incorporates an absorber and temperature transducer affixed to some mass with heat capacity $C(T)$. This mass is thermally linked to a stable temperature bath through thermal conductance $G(T)$. Initially, the entire device is in equilibrium with the bath at a temperature T_b. Radiation incident on the bolometer is absorbed, resulting in an increase of the temperature to T, which is detected by the thermometer. In this configuration the absorbed power is given by $P_{absorbed} = G(T)[T - T_b]$. The thermometer in the figure is typically some type of resistive material having a large temperature coefficient of resistance. In general, any element which yields a measurable quantity that is sensitive to changes in temperature may be used as the temperature transducer in the bolometer. In our case, the transducer is a temperature dependent inductor.

Figure 2 shows an example of a resistive bolometer with a very low electrical noise equivalent power NEP_e, fabricated by Downey *et al.* (1984). This bolometer is constructed of silicon using integrated circuit techniques. The center region is ion implanted in order to obtain a high temperature coefficient of resistance. A unique

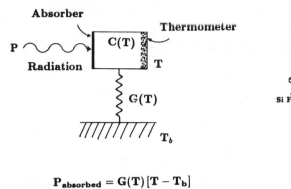

$$P_{absorbed} = G(T)[T - T_b]$$

Fig. 1 Thermal equivalent circuit for a bolometer.

Fig. 2 Illustration of front side of Downey bolometer.

feature of this device is its monolithic construction: the outer (mounting) frame, the thin Si threads for thermal isolation, and the temperature sensor are formed by etching a single piece of Si. The thermal conductance of this structure is 2.09×10^{-9} W/K at an operating temperature of 0.35 K. Its experimentally measured electrical noise equivalent power NEP_e is 6.1×10^{-16} W/$\sqrt{\text{Hz}}$. We will compare the theoretical performance of a superconducting bolometer to this device in the last section.

Another low NEP_e radiation detector is known as a superconducting transition edge bolometer. For this type of bolometer the temperature sensitive element is a superconducting film just above its transition temperature. A high quality device of this type was developed by Clarke *et al.* (1977). In their bolometer, an aluminum thin film on a sapphire substrate was suspended by indium coated nylon threads. These threads provided the electrical and thermal links to the device. The operating temperature is maintained close to the midpoint of the superconducting transition, shown in Fig. 3, where the film resistance changes rapidly with increasing temperature. The important point to notice here is that this is not literally a "superconducting" bolometer since the film is biased in the normal (resistive) state. Thus, a transition edge bolometer is subject to Johnson noise and self heating. The best NEP_e obtained by Clarke *et al.* (1977) with a bolometer of this type was 1.7×10^{-15} W/$\sqrt{\text{Hz}}$. Their device was operated at a temperature of 1.27 K and had an effective thermal conductance of 1.44×10^{-8} W/K.

3. SUPERCONDUCTING THERMOMETRY

Figure 4 depicts a superconducting stripline of width W, length l and thickness d_s, over a superconducting groundplane of width W_{gp} and thickness d_{gp}. The two films are separated by some dielectric of thickness t_0. An analysis of this system for $W >> t_0$ reveals that any low-frequency current in the microstrip line is essentially confined within a penetration depth λ in each of the superconducting films (Matick 1969). This current confinement is due to the Meissner effect, which implies that a magnetic field will exponentially decay to zero in the interior of a superconductor. The currents and magnetic fields are negligible at distances greater than λ from the film surface and hence are effectively confined within a superconducting penetration depth.

The inductance of the microstrip line shown in the figure is (Matick 1969):

$$L = \mu_0 \frac{l}{W}\left[t_0 + \lambda_s \coth\left(\frac{d_s}{\lambda_s}\right) + \lambda_{gp} \coth\left(\frac{d_{gp}}{\lambda_{gp}}\right)\right] \quad (1)$$

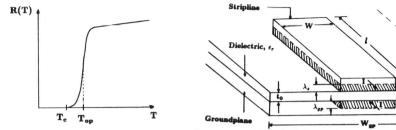

Fig. 3 Resistance versus temperature for a superconducting transition-edge bolometer.

Fig. 4 Schematic section of a superconducting microstrip line.

The first term is due to the magnetic field in the dielectric while the second and third terms are contributions from the two superconducting films. The important thing to notice here is that the inductance of the superconducting microstrip line structure has a functional dependence on the penetration depths of the two films, namely λ_s and λ_{gp}. While the penetration depth of a superconductor is analogous to the skin depth in normal metals, at audio frequencies the penetration depth is frequency independent and has only a very slight, and therefore usually negligible, dependence on an applied magnetic field. However, it has a strong dependence on temperature, for temperatures just less than T_c, as shown in Fig. 5, wherein

$$\lambda(T) = \frac{\lambda_0}{\sqrt{1 - (T/T_c)^4}} \tag{2}$$

Since the inductance of a superconducting microstrip line depends on the penetration depth, and hence upon temperature, one can construct a simple superconducting thermometer based upon this temperature dependence. A conceptually related device was constructed by Moody *et al.* (1984). Their device, depicted in Fig. 6, is a superconducting solenoid comprised of insulated niobium wire wrapped around a niobium core. This device, which we will refer to as the Moody thermometer, has a loop inductance L which is dependent upon the temperature. A persistent dc current I is induced in the superconducting loop and detected by an rf SQUID. A heater coil, wrapped around the bottom of the core, and germanium thermometer was used to establish the operating temperature of the device. In this device, a change in temperature about the bias point results in a change in the penetration depth $\lambda(T)$ and hence in the total inductance L. Since the total flux encompassed by a superconducting loop is quantized, (that is LI = const.), this results in a change in the circulating current I in the loop, to be detected by the amplifier. We can see in Fig. 5 that the most sensitive dependence of the penetration depth on temperature is near the critical temperature T_c and hence expect better sensitivity from this device in that region. The quoted temperature sensitivity δT was typically between 10^{-9} and 10^{-8} K/$\sqrt{\text{Hz}}$ (Moody *et al.* 1984), which is excellent thermometry. The drawback to a thermometer of this type is the 1/f noise of the SQUID amplifier, since it relies on a dc measurement.

4. KINETIC INDUCTANCE

A stronger dependence of the inductance on temperature can be obtained if one operates in the kinetic inductance limit. To understand this concept, consider an electric current in some conductor, with charge carriers e^*, having mass m, and number density n. The total energy due to this current can be expressed by

$$E = \int_{all\ space} \frac{1}{2}\mu_0 H^2 d\Omega + \int_{conductor} \frac{1}{2}mv^2 n d\Omega \equiv \frac{1}{2}LI^2 \tag{3}$$

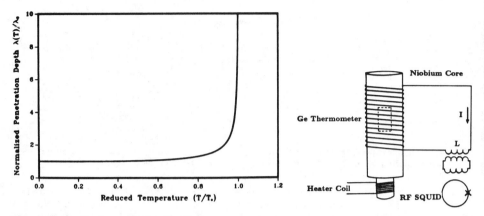

Fig. 5 Normalized penetration depth versus reduced temperature for a superconductor.

Fig. 6 Schematic of Moody penetration depth thermometer.

where $L = L_m + L_k$. The first term is the energy of the magnetic field over all space and is related to the "magnetic inductance" L_m while the second term is just the kinetic energy of the charge carriers in the conductor. The inductance associated with the second term is known as the "kinetic inductance" L_k. The kinetic inductance depends on the current distribution in the conductor and one can show for a homogeneous conductor of uniform cross section and current density that

$$L_k = \left(\frac{m}{ne^{*2}}\right)\left(\frac{l}{A}\right) \tag{4}$$

where l and A are the length and area, respectively, of the conductor. The first factor in Eq. 4 is dependent upon the material properties of the conductor while the second factor is geometry dependent.

It is now instructive to reconsider the microstrip line shown in Fig. 4, assuming equal thicknesses for identical stripline and groundplane superconductors (that is, $d_{gp} = d_s = d$; $\lambda_s = \lambda_{gp} = \lambda$). The total inductance of the microstrip line in this case is given by

$$L = \mu_0 \left(\frac{l}{W}\right)[t_0 + 2\lambda] \qquad (d >> \lambda) \tag{5a}$$

$$L = \mu_0 \left(\frac{l}{W}\right)\left[t_0 + 2\frac{\lambda^2}{d}\right] \qquad (d << \lambda) \tag{5b}$$

Equation 5a demonstrates that for a thick film geometry one has a linear dependence of the inductance on the penetration depth, as in the Moody thermometer. In the case of Eq. 5b, we find a stronger dependence on the penetration depth by a factor λ/d, where $\lambda/d >> 1$. Using the London expression for the penetration depth, $\lambda^2 = m/\mu_0 ne^{*2}$, we see that the second term in Eq. 5b has the form of the kinetic inductance given by Eq. 4. It is this second case which we shall refer to as the kinetic inductance limit. The present experiment is designed to be in this limit.

The kinetic inductance of a superconducting aluminum thin film was first measured by Little (1967). His experimental results, presented in Fig. 7, show the decrease in the dc resistance of the film to zero as the critical temperature T_c is approached.

Below T_c, the inductance of the film measured at each temperature decreases with decreasing temperature. The solid curve fit to his data is the kinetic inductance of Eq. 4, where $\lambda(T)$ is given by Eq. 2. This work experimentally established the idea of kinetic inductance.

5. KINETIC INDUCTANCE THERMOMETER

We are studying the kinetic inductance thermometer originally proposed by McDonald (1987) and shown in Fig. 8. If the four inductors of the figure are structurally identical, their inductances will be equal when they are all at the same temperature. In that case the bridge as a whole is not sensitive to temperature changes. However, the bridge is sensitive to temperature *differences* between the inductors. To make a practical differential thermometer, two of the inductors are weakly thermally coupled to the remainder of the bridge (assumed to be at the bath temperature). This feature ensures that the bridge can be balanced even with the inevitable fabrication variations among the inductors. As the power to the heaters is varied, the galvanometer in the figure senses the current null at bridge balance.

A SQUID is used as the null detector because it is the quietest of all amplifiers and can be impedance matched to the bridge. In fact, SQUID's can have nearly quantum limited sensitivity (Voss 1981, Koch *et al* 1981).

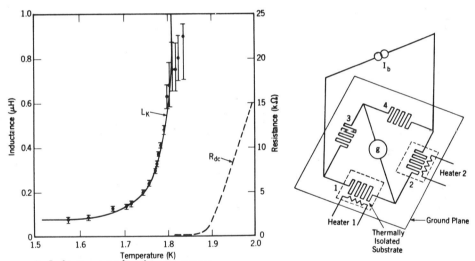

Fig. 7 Inductance and resistance versus temperature for an aluminum thin film $l = 3 \ cm; w = 20 \ \mu m; d = 4 \ nm$.

Fig. 8 Schematic of kinetic inductance thermometer.

Our present experiment has four meander line structures over a groundplane, each structure forming one arm of the bridge. Each meander line (see Fig. 9) consists of a thin film of niobium, 3 μm wide, 0.6 μm thick, 10 cm long, and fills a 1 mm^2 area. The figure also shows an AuIn$_2$ thin film wire to be used as the heater elements for the thermally isolated inductors. The Nb meander line is separated from the 0.02 μm thick Nb groundplane by 0.1 μm of Nb$_2$O$_5$ dielectric plus 0.02 μm of SiO dielectric. Figure 10 shows the front side of the integrated circuit bridge fabricated on a silicon substrate. Contact to the niobium wires is made through a 0.03 μm thick Au film on each niobium pad.

Two meander lines, labeled 1 and 2 in Fig. 8, are thermally isolated from the remaining circuit by a 15 μm thick membrane of boron doped silicon. This geometry is achieved by first patterning a boron doped frame around the meander line and then

Fig. 9 Photograph of niobium meander line inductor.

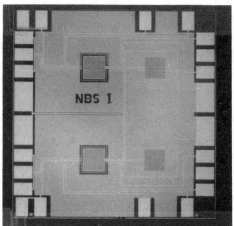

Fig. 10 Photograph of kinetic-inductance thermometer integrated circuit.

anisotropically etching the back side of the wafer, as shown in Fig. 11. This procedure leaves a silicon island suspended by a 55 μm wide boron doped silicon membrane. The estimated thermal conductance from this geometry is about 1×10^{-4} W/K. Figure 12 shows the back side of one of these silicon islands. The moat surrounding the island is defined by the etched silicon channels.

Fig. 11 Schematic of side profile of an etched silicon channel.

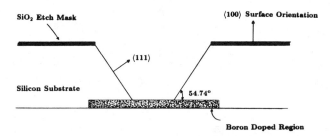

SiO$_2$ Etch Mask

(100) Surface Orientation

(111)

Silicon Substrate

54.74°

Boron Doped Region

Fig. 12 Photograph depicting etched silicon channels defining the back side of a thermally isolated silicon island.

The initial experiments are designed to compare the performance of the superconducting thermometer with an analytical model. This model predicts that the temperature sensitivity of the bridge circuit is inversely proportional to the bridge bias current. Hence an important question is the maximum allowed audio-frequency bias current I_b for the thermometer. Also, because the bridge is a differential thermometer, it should be capable of rejecting the residual temperature drifts and fluctuations of its thermal platform, which is important for cw measurements of power. This feature will also be examined experimentally.

We can obtain an estimate for the temperature sensitivity and electrical noise equivalent power for this kinetic inductance thermometer from the analysis outlined by McDonald (1987). This analysis predicts that the temperature sensitivity δT and noise equivalent electrical power NEP_e due to the amplifier are given by

$$\delta T = \frac{\delta I_g}{|dI_g/dT|}; \quad NEP_e = G\delta T \tag{6}$$

where

$$\frac{dI_g}{dT} = \left[\frac{-I_b/L_g}{4(1 + L_1/L_g)} \right] \left(\frac{dL_1}{dT} \right) \tag{7}$$

at bridge balance. The meander line inductance L_1 is formally given by Eq. 1, and in the kinetic inductance limit, is approximately given by $L_1 = \mu_0(l/W d_{gp})\lambda_{gp}^2$ for a thin dielectric, and $d_s \gg \lambda_s$; $d_{gp} \ll \lambda_{gp}$.

Assuming an operating temperature of $0.9T_c$, where T_c is 7.2 K for our Nb thin films, a bias current of 1 mA, and using a commercial dc SQUID having a typical noise-current of 1.5 pA/$\sqrt{Hz}$, the analysis predicts a δT of 0.2 $\mu K/\sqrt{Hz}$ and an NEP_e of approximately 2×10^{-11} W/$\sqrt{Hz}$ for our present thermometer design. This is already a order of magnitude better than the germanium resistance thermometer based detectors. Equations 6-7 imply that further improvements can be made by increasing the kinetic inductance L_1 of the meander lines. Substantial improvement in sensitivity can also be obtained by using an optimized SQUID amplifier (Clarke *et al.* 1979) and by decreasing the operating temperature of the device, as we shall now see.

It is instructive to compare the theoretical performance of a kinetic inductance bolometer with that of the experimental results of the silicon-based bolometer, discussed in Section 2, developed by Downey *et al.* (1984). The electrical noise equivalent power for the Downey bolometer is determined by the thermal fluctuations of the bolometer (phonon noise), Johnson noise, and amplifier noise and is given by

$$(NEP_e)^2 = (NEP_e)^2_{PHONON} + (NEP_e)^2_{JOHNSON} + (NEP_e)^2_{AMPLIFIER} \tag{8}$$

Their quoted NEP_e is 6.1×10^{-16} W/$\sqrt{Hz}$ (for $G = 2.09 \times 10^{-9}$ W/K), at 0.35 K, is predominantly from the latter two sources. Using the same thermal conductance and comparing the latter two noise terms for our superconducting bolometer coupled to an optimized SQUID amplifier (McDonald 1987), yields an NEP_e of 7×10^{-20} W/$\sqrt{Hz}$, at the same operating temperature as the Downey bolometer. This an improvement of about four orders of magnitude over the semiconductor device.

At 0.35 K the phonon noise term $(NEP_e)_{PHONON} = \sqrt{4k_B T^2 G}$ is 1.2×10^{-16} W/$\sqrt{Hz}$, which obviously would limit the device performance. Nevertheless, including phonon noise the superconducting device has about a factor of five improvement over the semiconductor device.

Another interesting comparison is with the sensitivity of devices designed to count optical photons. A high quality photomultiplier for 400 nm wavelength radiation has an NEP of 5.4×10^{-18} W/$\sqrt{\text{Hz}}$ when cooled to 150 K (Jones 1953). As noted above for the superconducting bolometer, the combined amplifier and Johnson noise can be below 1×10^{-19} W/$\sqrt{\text{Hz}}$, which is well below the photomultiplier noise. Consequently with the superconducting bolometer we are concerned only with the phonon noise. For a practical design we use data from the Downey bolometer, for which the thermal conductance is given by $G = G_0 T^3$, where G_0 is 4.88×10^{-8} W/K^4. Using the usual formula for the phonon noise we can calculate the operating temperature required to produce a phonon noise equal to the NEP of the photomultiplier. That temperature is 0.1 K, which is a practical value. We conclude that a superconducting bolometer potentially has a sensitivity comparable to that of a photon counting device.

6. SUMMARY

We are presently developing a temperature transducer which is effectively free of Johnson noise because it is an all superconducting device. In its ideal configuration the bridge circuit, with inductive elements composed of superconducting meander lines in the kinetic inductance limit, is coupled to a nearly quantum limited SQUID amplifier used as the null detector. It should be mentioned that this type of device can be directly calibrated by electrical substitution.

7. ACKNOWLEDGEMENTS

It is a pleasure to acknowledge contributions to this work by R.J. Phelan, W.A. Wichart, R.H. Ono, J.E. Beall, and F.L. Lloyd. Partial financial support has been provided by the U.S. Army, Redstone Arsenal, through the office of M. Fecteau.

8. REFERENCES

Clarke J, Hoffer G I, Richards P L and Yeh N H 1977 *J. Appl. Phys.* **48** 4865

Clarke J, Tesche C D and Giffard R P 1979 *J. Low Temp. Phys.* **37** 405

Downey P M, Jeffries A D, Meyer S S, Weiss R, Bachner F J, Donnelly J P, Lindley W T, Mountain R W and Silversmith D J 1984 *Appl. Opt.* **23** 910

Jones R Clarke 1953 *Advances in Electronics* ed L Marton (New York: Academic Press) 1

Koch R H, VanHarlingen D J and Clarke J 1981 *Appl. Phys. Lett.* **38** 380

Little W A 1967 *Proceedings of the Symposium on the Physics of Superconducting Devices* Charlottesville, Virginia pp S-1 to S-17

Matick R E 1969 *Transmission Lines for Digital and Communication Networks* (New York: McGraw-Hill) Chap.6

McDonald D G 1987 *Appl. Phys. Lett.* **50** 775

Moody R E, Chan H A, Paik H J and Stephens C 1984 *Proc. 17th Int. Conf. on Low Temp. Phys.* ed U Eckern, A Schmid, W Weber and H Wuhl (Amsterdam: North-Holland) pp 407-8

Voss R F 1981 *Appl. Phys. Lett.* **38** 182

Inst. Phys. Conf. Ser. No. 92
Paper presented at Int. Conf. Optical Radiometry, NPL, London, 12–13 April 1988

Characteristics of Ge and InGaAs photodiodes

Edward Zalewski

National Bureau of Standards (US), Gaithersburg, MD 20899

ABSTRACT: Measurements of the internal quantum efficiency of recently developed near ir (1 to 1.6 microns) photodiodes show that considerable improvement has been made in the radiometric quality of these devices. Among commercially available devices, the newer InGaAs/InP photodiodes exhibit better characteristics than the Ge devices that have been traditionally used for near ir radiometry. However, experimental induced junction Ge photodiodes produced at Purdue University have been observed to have nearly ideal internal quantum efficiency.

1. INTRODUCTION

Recent developments in fiber optics for the near ir have placed new demands on radiometric accuracy and photodetector quality in this spectral region. This, in turn, has led to the development of a new type of photodetector for the 1 to 1.6 micron region, an InGaAs photodiode, and has stimulated improvement of the long-standing detector of choice for this region, the Ge photodiode. The radiometric quality of commercially available photodiodes of both types has undergone substantial improvement in just the last two years. Furthermore, the prognosis for even more improvement in the very near future is excellent.

The internal quantum efficiency of several of both types of presently available photodiodes has been measured at three laser wavelengths in the near ir. The results of these measurements are discussed in relation to the general features of the construction of the photodiodes and the implications regarding the origin of the internal quantum efficiency losses within the different types of devices. Besides being a good indicator of the radiometric quality of a particular device, a nearly ideal internal quantum efficiency implies that the device may be suitable for self-calibration in a manner similar to that of a high-quality silicon photodiode (Zalewski and Geist, 1980; and Geist et al, 1980).

2. EXPERIMENTAL PROCEDURES

All the photodiodes examined in this study had their windows removed to enable a more accurate measure of the reflectance loss at the front surface. Three types of InGaAs photodiodes and one type of Ge were examined. One type of InGaAs was an epitaxially grown "mesa type" with an InP cap purchased in 1986 from the Epitaxx Company. The other two types of InGaAs photodiodes were of the newly introduced planar diffusion construction by Epitaxx, ~~one type with and one without the InP~~ cap. The Ge photodiodes were very recently purchased from the EG&G Judson Company. [These commercial products are identified for the sole purpose of adequately describing the experimental results presented in this study. In no event does such identification imply recommendation or endorsement by the National Bureau of Standards.]

The measurements were made using three laser wavelengths: HeNe at 1.15 and 1.52 microns, and a laser diode at 1.32 microns. The configuration of the experimental instrumentation followed well established electro-optical radiometry methods first described by Geist et al (1975, 1977). The unique component was the liquid crystal modulator system used to stabilize the laser power. This was a prototype instrument developed by Peter Miller of the Cambridge Research Instrumentation Company (USA). An InGaAs/InP photodiode was used with a wedged beam splitter to sample the laser power for feedback control via the electrically adjustable birefringence of the liquid crystal.

The absolute spectral response (external quantum efficiency) was measured by reference to an electrically calibrated pyroelectric radiometer (ECPR). The absolute calibration of the ECPR was obtained by comparison to an absolute silicon photodiode detector (Zalewski and Duda, 1983) using visible laser radiation. The visible to near infrared reflectance of the ECPR was constant to within 0.1 % as measured by the manufacturer (Laser Precision Company) and experimentally verifed by comparison to a cavity type thermal radiometer at 0.6 and 1.3 microns. The estimated total uncertainty in the external quantum efficiency measurement was approximately 0.5 %.

The internal quantum efficiency was calculated by using the value of the reflectance measured simultaneuosly with the absolute response. Simultaneous measurement of the reflectance and response assures that the angular and spatial nonuniformities of the photodiodes will not degrade the accuracy of the internal quantum efficiency calculation (see Eppeldauer et al, 1987).

The reflectance was measured by using one of the other detectors in the group as the reflectometer. The distance between the test detector and the reflectometer was kept as small as possible (about 4 cm). This meant that the measured reflected radiation was mostly specular (within a 2 degree right circular cone). The InGaAs photodiodes had optical quality surfaces so that the diffuse component of the reflection could be assumed to be very small relative to the specular. The Ge photodiodes surfaces, however, had a slight "orange peel" appearance and could be expected to have a larger diffuse component. One experiment at a larger distance (sampling within an approximately 1 degree right circular cone) indicated a negligible decrease in reflectance for the InGaAs but a non-negligible (about 0.003 out of 0.13) decrease for the Ge. A

more complete study of the reflectance properties of the
various detectors used in this study would, of course, provide
a more accurate assessment of the internal quantum efficiency
of each type of device. As will be seen below, this is not a
critical point at this time, since the devices with the nearly
ideal internal quantum efficiency also have the best optical
quality reflectance.

3. RESULTS

The measured internal quantum efficiency of the various types
of photodiodes discussed above are shown in Figs. 1 through 4.
The internal quantum efficiency of two mesa type InGaAs/InP
(with an InP cap) photodiodes is presented in Fig. 1. These are
the devices studied by Eppeldauer et al (1987) at 1.32 microns.
The data from that study and the present one are in excellent
agreement. The lines drawn on the graph connect the average of
each set of measurements, showing the remarkable simularity
between these two devices and the closeness to the ideal of
100% quantum efficiency at 1.15 microns. At this wavelength,
the measured internal quantum efficiency for the two devices
was 0.998+0.006 and 0.993+0.008, respectively. The uncertainty
here is the overall absolute (one sigma) uncertainty for each
measured value. The specular reflectance at this wavelength was
quite low (about 0.04). The surface, as noted above, had an
optical quality appearance so that the diffuse reflectance was
a negligible component of the absolute uncertainty of this
measurement.

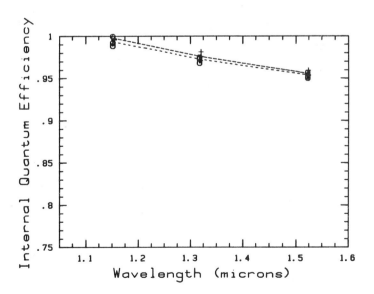

Fig. 1 Mesa type InGaAs/InP photodiodes, two devices
measured.

Figures 2 and 3 show the measured internal quantum efficiency of the other two types of InGaAs photodiodes studied. Both are of the planar diffusion construction; the two devices depicted in Fig. 2 have an antireflection coating and the normal InP cap, whereas the device depicted in Fig. 3 has no anti-reflection coating and no InP cap. For comparison with the devices depicted in Fig. 1, at 1.15 microns the devices depicted in Fig. 2 have specular reflectances of about 0.02 while that of Fig. 3 is 0.32. The greater spread of the data in Fig. 3 is due to the variability of the measurement of the larger reflectance loss. The other major difference between the various InGaAs devices studied is most likely the depth of the junction in the InGaAs layer. In the mesa technique, the zinc dopant is introduced into the furnace during the final stage of the epitaxial growth of the InGaAs layer, thereby keeping it at the InP/InGaAs interface. In the planar diffusion technique the zinc dopant apparently penetrates deeper, since these devices exhibit a greater capacitance. From the observed polarities of the photocurrents, the InGaAs photodiodes were p on n and the Ge photodiodes (Fig. 4) were n on p.

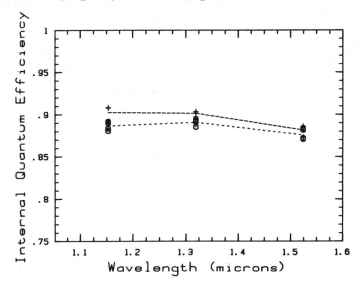

Fig. 2 Planar diffused type InGaAs/InP photodiodes, two devices measured.

Figure 4 shows the internal quantum efficiency of two Ge photodiodes. These data are in good agreement with the internal quantum efficiencies reported by Stock (1987) on his most recently procurred Ge photodiodes. The internal quantum efficiency of Ge as measured here is qualitively the same as that measured by Nofziger on experimental np planar diffused devices designed and prepared by Huang and Schwartz (1988). It is interesting to note that also they reported nearly ideal internal quantum efficiency on some induced junction experimental photodiodes over the spectral range from 1 to 1.6 microns.

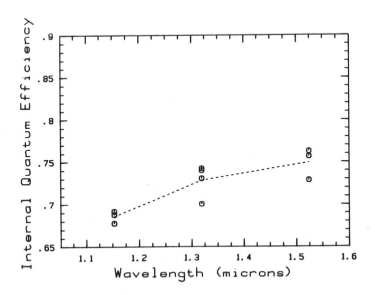

Fig. 3 Planar diffused type InGaAs photodiode, no antireflection coating and no InP cap.

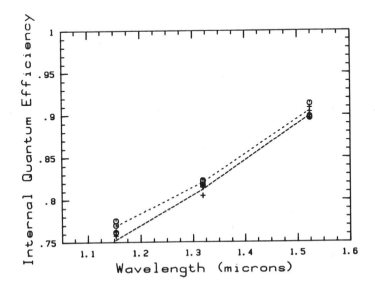

Fig. 4 Ge photodiodes, two devices measured.

4. DISCUSSION AND CONCLUSIONS

The Ge photodiodes are constructed of p type base material into which is diffused a shallow n type junction. The calculations of Huang and Schwartz (1988) show that shallower junctions lead to higher internal quantum efficiencies of the shape that we observe here. They also show that nearly ideal internal quantum efficiencies cannot be realized unless the surface recombination velocity is substantially reduced, that is, unless the surface of the Ge is well passivated. Unlike silicon, the oxide of germanium, while capable of producing a well passivated surface, cannot maintain it because it is water soluble. Ge photodiodes with a surface coating of silicon dioxide (Huang and Schwartz, 1988) or germanium oxynitride (Hymes, 1988) will have improved stability and a reasonable level of surface passivation. Since we do not observe an ideal internal quantum efficiency in the Ge photodiodes we have measured, we conclude that there is a rather high degree of surface recombination taking place. This conclusion is qualitatively substantiated in Fig. 5. Photons penetrating to a greater depth produce charge carriers that are collected with a greater efficiency.

The penetration depth of photons at the three wavelengths studied is plotted in Fig. 5. The solid curves indicate the fraction of photons absorbed as a function of depth in Ge. The absorption coefficient decreases as a function of wavelength, so that the 1.15 micron radiation at a depth of 8 microns is

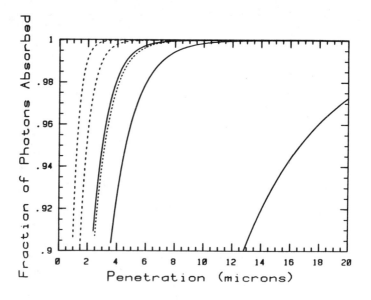

Fig. 5 Photon penetration depth in InGaAs (dashed curves) and Ge (solid curves), see text.

nearly 100% absorbed. The values of the absorption coefficient used to obtain these curves are those of Dash and Newman (1955): 1.0/micron at a wavelength of 1.15 microns, 0.65/micron at 1.32 microns, and 0.18/micron at 1.52 microns. The dashed curves indicate the fraction of photons absorbed as a function of depth in InGaAs. The values of the InGaAs absorption coefficient are those of Weidner and Geist (private communication): 2.37/micron at a wavelength of 1.15 microns, 1.55/micron at 1.32 microns, and 0.950/micron at 1.52 microns.

The construction of an InGaAs photodiode begins with a base of highly n doped InP as a substrate onto which is grown a thin layer of weakly n doped InP that acts as a buffer (Olsen, 1985). The InGaAs layer is then epitaxially grown on the InP substrate. It is in this layer that the junction is grown by either of the two techniques described in the previous section. The InP cap on top of the InGaAs passivates the interface, however, it is opaque to visible radiation making the device inoperable in this spectral region. The final layer is an antireflection coating of a transparent material such as silicon nitride.

Referring to Figs. 1 and 5, we can see that for the mesa device the internal quantum efficiency is highest for photons having the smallest penetration and drops by 5% for photons that penetrate twice as far. In Fig. 2 we see that the internal quantum efficiency is down by about 10% from that of Fig. 1 and that the effect of penetration depth is less pronounced. Finally in Fig. 3 we see an even greater drop in internal quantum efficiency with a spectral dependence somewhat like that of the Ge photodiodes. A qualititative explanation of these observations is that for the mesa grown device good passivation and a very shallow junction are obtainable, whereas that is not the case in the planar diffuse devices. It was expected that for the mesa devices the application of a reverse bias voltage, which causes an increase in the depth of the charge depletion region, would produce an increased quantum efficiency at longer wavelengths. Unfortunately it was found that the mesa devices cannot withstand the bias voltage necessary to produce a discernible effect. The thickness of the InGaAs layer was not known, however, if the thickness of the InGaAs is about 3 microns as described by Olsen (1985, Fig. 3a), then the internal quantum efficiency is approximately equal to the fraction of photons absorbed in this layer.

The reason for adopting the planar diffuse technology to InGaAs manufacture is to produce larger area devices with greater stability. In the mesa device the edge of the junction extends to the edge of the wafer and is therefore not protected. In many applications trading high quantum efficiency for stability is very desirable. As the understanding of the effect of the construction parameters on the internal quantum efficiency is increased, the various options will become more obvious and we may expect to see substantial improvements in the radiometric characteristics of both the InGaAs and Ge devices.

5. ACKNOWLEDGEMENTS

The author wishes to expess his gratitude to: R. Brubaker, G. Eppeldauer, J. Geist and D. Thomas of NBS; P. Miller of the Cambridge Research Instrumentation Company; G. Gasparian and G. Olsen of the Epitaxx Company; and to T. Wong of the EG&G Judson Infrared Company.

6. REFERENCES

Dash W C and Newman R 1955 Phys. Rev. Vol 99 p 1151

Eppeldauer G, Tsudagawa M, Zalewski E and Houston J 1987 Proc. 13th IMEKO Int. Symposium on Photon Detectors (IMEKO, Braunschweig) in press

Huang D L and Schwartz R J 1988 " Design and Fabrication of a Self-Calibrating Germanium Photodiode for Radiometric Applications" Report to NBS No. TR EE 88 4 (Purdue University, West Lafayette, IN); see also Schwartz R J and Huang D L 1984 Proc. SPIE Vol 499 p 2

Hymes D J 1988 "Growth, Physical Properties and MOS Characterization of Native Germanium Oxynitride Films" Thesis (Brown University, Providence, RI)

Geist J, Steiner B W, Schaefer A R, Zalewski E F and Corrons A 1975 Appl. Phys. Lett. Vol 26 p 309

Geist J, Lind M A, Schaefer A R and Zalewski E F 1977 "Spectral Radiometry: A New Approach Based on Electro-Optics" NBS (USA) Tech Note 954 (US Government Printing Office; Washington DC 20402)

Geist J, Zalewski E F and Schaefer A R 1980 Appl. Optics Vol 19 p 3795

Olsen G H 1985 "Long Wavelength Components by Vapor Phase Epitaxy" Laser Focus/Electro-Optics Vol 12 Jan. p 124

Stock K D 1987 Appl. Optics Vol 27 p 12

Zalewski E F and Geist J 1980 Appl. Optics Vol 19 p 1214

Zalewski E F and Duda C R 1983 Appl. Optics Vol 22 p 2867

Inst. Phys. Conf. Ser. No. 92
Paper presented at Int. Conf. Optical Radiometry, NPL, London, 12–13 April 1988

Semiconductor photodiodes as detectors in the VUV and soft x-ray range

E. Tegeler
Physikalisch - Technische Bundesanstalt, Abbestr. 2-12, D-1000 Berlin 10, Germany

M. Krumrey
BESSY GmbH, Lentzeallee 100, D-1000 Berlin 33, Germany

ABSTRACT: The properties of semiconductor photodiodes originally designed for the visible spectral range have been investigated at photon energies from 6 eV to 3500 eV. For silicon photodiodes a strong radiation induced decrease of the responsivity was found. The spectral dependence of these effects has been studied in the vicinity of the Si L absorption edge. GaAsP- and GaP- Schottky diodes show remarkable stability and high quantum efficiency.

1. INTRODUCTION

Radiation in the spectral range of the Vacuum Ultra-Violet (VUV) and the soft X-rays is of growing importance in many fields of basic research (atomic, surface, solid state, and astrophysics), applied research (lasers, thermonuclear fusion, material research), and manufacturing (lithography for microcircuit engineering and micromechanics). For these applications easy to operate detectors of known spectral responsivity, socalled transfer standards, are required.

The most common detectors of known spectral responsivity in the VUV and soft X-ray range are photoemissive detectors, in a large spectral region preferably a gold photocathode, utilizing published data for the quantum efficiency of Au. The quantum efficiency of gold was determined by comparison with rare gas ionisation detectors (ionisation chamber and proportional counter), the quantum efficiency of which can be calculated. Unfortunately the published data for the quantum efficiency of Au differ by up to 50% (Henke et al 1981, Day et al 1981). Furthermore photocathodes are extremely sensitive to surface contaminations.

For the establishment of a responsivity scale in the VUV and soft X-ray region therefore two problems have to be solved:
 the development of detectors that can be used as reliable transfer standards
 the development of a calibration procedure for the precise determination of the spectral responsivity of detectors.

This paper will deal only with the suitability of detectors as secondary standards. For this purpose also spectral responsivities will be presented, but these responsivities have been measured by comparison with photoemissive detectors and do not fullfill radiometric requirements.

The most reliable standard detectors from the infrared to the ultraviolet and also in the X-ray region are semiconductor detectors. Their properties are very promising also in the VUV spectral range, because they are compatible to ultra high vacuum conditions, relatively insensitive to surface contaminations, easy to handle, small, and inexpensive. Until recently absorbing deadlayers have prevented their use in the VUV, but improved production methods resulting in thinner deadlayers have raised the possibility to utilize such diodes in this spectral region.

2. EXPERIMENTAL

The semiconductor diodes investigated here are commercially available photodiodes where the glass or silica window has been removed from the housing. They are operated in the photoamperic mode (Geist 1986), usually without bias. The photocurrents, which are in the range from a few pA up to hundreds of nA were measured using a picoammeter (Keithley Electrometer 617).

As the data presented in this paper cover the spectral range from 6 eV to 3.5 keV, we refer to measurements which has been carried out in collaboration with other groups at several monochromators at the synchrotron radiation centers BESSY and HASYLAB (Barth et al. 1986, Krumrey et al. 1988). For a first determination of their spectral responsivity, the semiconductor diodes were compared to the usual standard detectors : A CsTe-diode with MgF_2-window in the spectral range 6 eV-10 eV and a Al_2O_3-diode in the spectral range 10 eV - 40 eV; both diodes have been provided by the NBS. For photon energies E > 40 eV a gold diode was used, which was produced by in situ evaporating 0.1 μm Au on a stainless steel surface with a ring anode installed in front of the photocathode to remove the emitted photoelectrons. For the determination of the quantum efficiency, yield data of Henke et al.(1981) were used combined with measurements of Lenth (1979) and cross section data of Veigele (1973).

The investgations on the stability of the diodes have been conducted at the Toroidal Grating Monochromator (TGM) in the laboratory of PTB at BESSY (Fischer et al. 1984). At this monochromator the flux of monochromatic radiation behind the exit slit has been shown to be proportional to the electron current in the storage ring. Relative changes in the responsivity of detectors therefore can be measured with an uncertainty of < 0.1%.

3. CHARACTERISTICS OF SEMICONDUCTOR DIODES IN THE VUV AND SOFT X-RAY RANGE

Two types of semiconductor photodiodes are commonly used as detectors with thin deadlayers (socalled UV enhanced diodes): diffusion type diodes and Schottky diodes. In the diffusion type diode a depletion layer is formed between the p - and n - doped layers, and in a Schottky diode at the semiconductor - metal interface.

A photon, which is absorbed in the semiconductor, will produce a number of electron-hole pairs. The mean energy w required for the creation of one electron-hole pair depends on the band gap of the semiconductor (Alig et al. 1980). Experimental values for w are 3.61 eV for Si, 4.2 eV for GaAs (Sakai 1982) and 6.54 eV for GaP (Kobayashi 1972) . If these pairs are created in the depletion layer or if they diffuse into this space charge region, they are separated by the internal electric field giving rise to an external photocurrent. Therefore the spectral responsivity (i.e. the the photocurrent per incident radiation power) can be described by

$$s(E) = (e/w) \cdot C \cdot \tau_1 \cdot (1-\tau_2)$$

where τ_1 and τ_2 denote the transmittance through the surface deadlayer (τ_1) and the space charge region (τ_2), C the (spectrally dependent) collection efficiency for the photogenerated carriers, e/w is the maximum possible responsivity for an ideal diode.

On the basis of this model we will discuss the spectral responsivities of GaAsP-Au and GaP-Au Schottky diodes, which were measured by comparison with photoemissive diodes (see Fig. 1). The theoretical maximum responsivities e/w are 0.153 A/W for the GaP- and 0.182 A/W for the GaAsP-diode (using with Wilson and Lyall (1986) a composition GaAs$_{0.63}$P$_{0.37}$). The measured maximum responsivities are close to these values. The overall spectral dependence of the reponsivities is very similar for the two types of diodes. At low photon energies the responsivity is strongly limited by the small transmittance τ_1 of the evaporated gold film (thickness 10 nm) which acts as a deadlayer. The gold transmittance also accounts for the valley between 150 eV and 500 eV, while the loss in responsivity at higher energies originates from the penetration of radiation through the depletion layer as described by the term $(1-\tau_2)$. In the high energy region an increase of the absorptance of the semiconductor material therefore will result in an increase of responsivity, as observed at the Ga L$_3$ absorption edge (1116 eV) and the P K absorption edge (2149 eV).

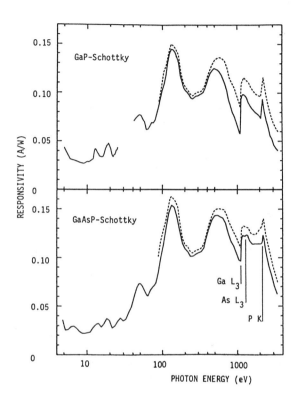

Fig. 1: Responsivity of a GaP- and a GaAsP-Schottky diode without bias (———) and with a bias of 4.5 V (------)

By applying a bias of 4.5 V (dashed curves in Fig. 1) the responsivity is increased in the low energy range by at most 10% due to the improved charge collection efficiency C. At high photon energies the effect of a bias voltage is much more pronounced. The observed behaviour can be explained by the increase of the depletion layer caused by the bias as predicted by the theory of Schottky contacts (Bertolini and Coche1968).

From their spectral responsivity semiconductor photodiodes are well suited as standard detector: contrary to photoemissive diodes the reponsivity remains relatively high up to photon energies of several keV. With the exception of some absorption edges in the region above 1 keV no sharp spectral features are observed. From theory and in accordance with our experience the semiconductor diodes are relatively insensitive to surface contaminations, because the creation of the charge carriers is a bulk effect and not a surface related effect as in the case of photoemissive diodes. Contaminations will act only as an additional deadlayer with a thickness of a few monolayer.

4. STABILITY OF DIFFUSION TYPE SILICON DIODES

For the suitability as a standard detector a very important feature is the stability of the responsivity. From the ultraviolet spectral range it is known that Si diodes show considerable aging effects (Korde and Geist 1987). As this behaviour is closely related to the photoabsorption cross section, we have studied the properties of different types of diodes in the vicinity of the Si L absorption edge (90 eV - 200 eV).

The types of diodes that were investigated are listed in Tab. 1. When irradiated with intense VUV radiation all Si- diodes show a decrease in responsivity at a typical time scale of the order of a few minutes. After the irradiation the diodes show regeneration effects with time constants of some months.

An extraordinary behaviour was found with the inverted channel type diode: this diode shows a decrease in responsivity by more than a factor of 100 within a few seconds when irradiated with the intense zero order radiation of the TGM. The responsivity could partly be recovered when irradiating not only the center of the diode, but also the areas near the boundary of the sensitive surface.

As an example of a "typical" behaviour we discuss the aging of a pnn$^+$-diode at a photon energy of 110.7 eV. Fig. 2 shows the decrease of the spectral responsivity when the diode is irradiated with app. $3 \cdot 10^9$ photons/s into an area of appr. 1 mm^2.

Table 1: Semiconductor photodiodes studied in the VUV

Photodiode Code No.	type	sensitive area (mm^2)	remarks
Hamamatsu S1722	Si - pin	17	
EG&G UV-444BQ	Si - pin	100	
Hamamatsu S1226-8BQ	Si - pnn$^+$	35	
Hamamatsu S1226-8BQ without SiO2-coating	Si - pnn$^+$	35	slow aging
EG&G BA17-100-100	Si - pdsb*)	100	
UDT-UV100L (inverted channel)	oxide-n$^+$p	100	rapid aging
UDT-pin10	Si-Schottky	10	
Hamamatsu G1127-02	GaAsP-Schottky	21	stable
Hamamatsu G2119	GaAsP-Schottky	100	electronic problems
Hamamatsu G1963	GaP-Schottky	21	stable

*) partially depleted silicon surface barrier detector, windowless, in general used as charged particle detector

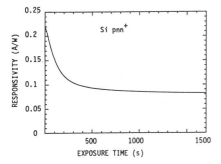

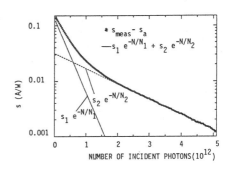

Fig. 2. Decrease of the responsivity of a pnn$^+$ diode when irradiated with $3 \cdot 10^9$ photons/s into an area of appr. 1 mm^2 at a photon energy E = 110.7 eV.

Fig. 3. The asymptotic behaviour of the decrease of the responsivity as shown in Fig. 2 can be described by $s = s_a + s_1 \exp(-N/N_1) + s_2 \exp(-N/N_2)$ with $s_a = 0.0813$ A/W, $s_1 = 0.1326$ A/W, $N_1 = 3.27 \cdot 10^{11}$ photons/s, $N_2 = 1.57 \cdot 10^{12}$ photons/s.

From Fig. 2 a saturation behaviour of the aging during the irradiation can be expected, in this particular case with an asymptotic reponsivity of $s_a = 0.0813$ A/W. In Fig 3 the deviation from the asymptotic responsivity $(s - s_a)$ is shown as a function of the number of photons deposited on the photodiode. The figure suggests to describe the the aging with two exponential decreases:

$$s = s_a + s_1 e^{-N/N_1} + s_2 e^{-N/N_2}$$

For measurements at different photon energies the aging of the diodes can very precisely be approximated with this mathematical expression (see Fig. 3). In order to understand the aging effect of the diode it is useful to investigate its spectral dependence for the aging constants $1/N_1$ and $1/N_2$ as well as for the responsivities s_a and $s_n = s_a + s_1 + s_2$.
We have measured the aging of the diodes for more than 15 specimens of the same type of pnn$^+$ diode. For diodes which had each been aged at a different photon energy the spectral dependence of the aging constants is shown in Fig. 4. Both aging constants $1/N_1$ and $1/N_2$ show an increase at a photon energy of 107 eV. In the lower part of Fig. 4 the absorption constants of Si (Gahwiller and Brown 1970, Brown and Rustgi 1972) and SiO$_2$ (Palik 1985) are shown. The increase of the aging constants obviously is related to the increase of the absorptance of SiO$_2$, while the increase the absorption edge of pure Si has no influence on the aging constants.

Until now we have no model for the existence of two aging constants $1/N_1$ and $1/N_2$. Eventually inhomogenities in the irradiation of the diode could be the reason for the formal description with two aging constants.

The measured spectral responsivities s_n for a "new" diode and s_a for an "aged" diode are shown in the upper part of Fig. 5. The spectral responsivity of a "new" diode is close to the value of $s_{max} = 0.278$ A/W = e/3.6 eV. The spectral responsivity of the "aged" diode s_a is drastically modified, and the dominating feature (besides the SiO$_2$ absorption edge at 107 eV which already appears in the spectrum of the "new" diode) is the Si L absorption edge at 99 eV.

The relation between the aging effect and the absorption of silicon is shown in the bottom part of Fig. 5: The loss in responsivity, i.e. the ratio s_a/s_n, shows a close similarity to the transmittance loss which has been calculated with the help of published absorption data for a Si-layer of 48 nm thickness (Gahwiller and Brown 1970, Brown and Rustgi 1972).

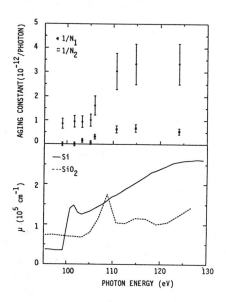

Fig. 4:
upper part: Aging constants $1/N_1$ and $1/N_2$ as a function of photon energy.
bottom part: absoption of Si and SiO_2. The spectral dependence of the aging constants is correlated to the absorption of SiO_2.

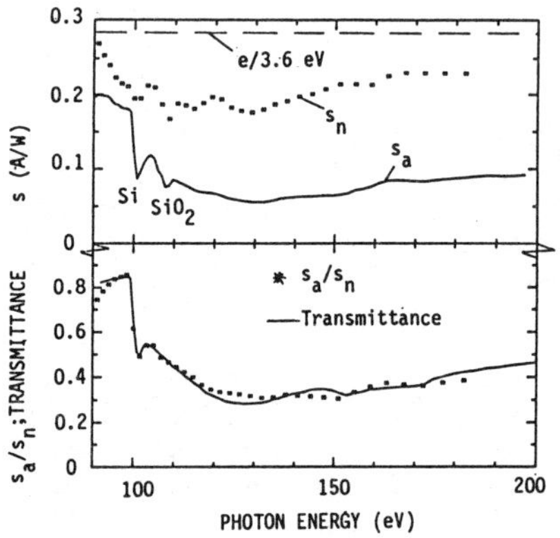

Fig. 5:
upper part: responsivity for a "new" and an "aged" Si diode.
bottom part: Ratio s_a/s_n for the responsivities of an "aged" and a "new" diode compared to the transmittance of a layer of 48 nm Si.

We summarize the aging behaviour of the Si-pnn$^+$-diode:
In the spectral range around the Si L absortion edge aging of Si-diodes is caused by absorption of photons in the SiO_2 passivation layer. The effect of the aging can formally be described by the formation of a deadlayer of Si. The formation of the Si deadlayer shows an asymptotic thickness of 48 nm. The diodes partly recover if they are not irradiated. The unstable deadlayer of Si prevent the use of duffusion type Si diodes at least in the spectral range 10 eV - 700 eV.

As the aging constants are correlated with the absorption in the insulating SiO_2 layer, better results are expected for diodes which do not have an insulating layer near the surface.

5. STABILITY OF SCHOTTKY DIODES (GaAsP AND GaP)

Contrary to the diffusion type diodes Schottky diodes do not have an insulating passivation layer. We therefore started an investiation with commercially available

Schottky diodes. With a Si Schottky diode we found aging effects similar to that of the pnn⁺-diode.

GaAsP- and GaP-Schottky diodes (see Tab. 1) have been shown to be excellent detectors in the UV even for radiometric purposes (Wilson and Lyall 1986). Because of their thin coating of only 10 nm Au they have low losses in surface deadlayers. Moreover, III-V compounds are known to be more resistant to radiation damage than Si.

The measurement of the aging effect for GaAsP-Au and GaP-Au Schottky diodes were performed in the same way as the test of the Si diodes (see Fig. 2) and the result is shown in Fig. 6. Within the uncertainty of <0.1 % the diodes show no aging effect even when irradiated with the intense zero order radiation of the Toroidal Grating Monochromator.

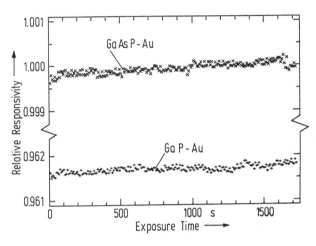

Fig. 6: Stability of the responsivity of a GasP and a GaP Schottky diode when irradiated with 10¹⁰ photons/s at a photon energy E = 124 eV

We have investigated about 20 specimens of the GaAsP diode and 15 specimens of the GaP diode. Except for one GaAsP diode which obviously had a carbon containing coating all diodes show very similar behaviour and the responsivity of all specimens of the same type of diode is equal within 10% in the entire spectral region.

The Schottky diodes can be used through the entire spectral region of the UV, the VUV and the soft X-ray range. Considering also the spectral responsivity presented in Fig. 1, we have to conclude that GaAsP and GaP Schottky diodes show very promising characteristics for the use as secondary standard detector.

6. CONCLUSION AND OUTLOOK

With GaAsP and GaP Schottky diodes reliable secondary standards are available. The spectral responsivities presented in this paper are measured by comparison with photoemissive diodes and therefore do not fullfill radiometric requirements. For the establishment of a responsivity scale reliable calibration precedures have to be developed. Rare gas double ionisation chambers which are used as primary detector standards for photon energies below 100 eV cannot be utilized at energies of several hundreds eV. Thus alternative calibration procedures have to be developed.

A responsivity scale can be based either on primary detector standards or on primary radiation standards. The electron storage ring BESSY is a standard radiation source with uncertainties between 0.23% in the visible and 2% in the the soft X-ray region (Riehle and Wende 1987). PTB therefore has started a program for the development of a calibration procedure which is based on the characteristics of the primary radiation standard BESSY.

Acknowledgements

We thank J. Barth, M. Krisch, U. Kroth, M. Kühne, P. Müller F. Schäfers, R. Thornagel, G. Ulm, B. Wende, and R. Wolf for experimental support and helpful discussions.

References

Alig R C, Bloom S, and Struck C W 1980, Phys. Rev. B **22**, 5565
Barth J, Tegeler E, Krisch M, and Wolf R 1986, in "Soft X-Ray Optics and Technology", E.E. Koch and G. Schmahl, editors, Proc. SPIE **733**, 481
Bertolini G and Coche A 1968, North Holland, Amsterdam, p. 133
Brown F C and Rustgi O P 1972, Phys. Rev. Lett. **28** 497
Day R H, Lee P, Saloman E B, and Nagel D J 1981, Report Los Alomos LA-012-79-1360
Fischer J, Kühne M, and Wende B 1984, Appl. Opt. **23**, 179
Gahwiller C and Brown F C 1970, Phys. Rev. **B2** 1918
Geist J 1986, Appl. Opt. **25**, 2033
Henke B L, Knauer J P, and Premaratne K 1981, J. Appl. Phys. **52**, 1509
Henke B L, Lee P, Tanaka T J, Shimabukuro R L, and Fujikawa B K 1982, Atomic Data and Nuclear Data Tables **27**, 1
Kobayashi T 1972, Appl. Phys. Lett. **21**, 150
Korde R and Geist J 1987, Appl. Opt. **26**, 5284
Krumrey M, Tegeler E, Barth J, Kirsch M, Schäfers F, and Wolf R 1988, Submitted to Appl.Opt.
Lenth W, Thesis, University of Hamburg, published in "Instrumentation for Spectroscopy and other Applications", in "Synchrotron Radiation", C. Kunz, editor, (Springer, Berlin 1979)
Palik E D 1985, Handbook of Optical Constants of Solids,New York
Riehle F and Wende B 1987, Optik **75**, 142
Sakai E 1982, Nucl. Instr. and Meth. **196**, 121
Veigele W M 1973, Atomic Data Tables **5**, 51
Wilson A D and Lyall H 1986, part 1 and 2, Appl. Opt. **25**, 4530 and 4540

Inst. Phys. Conf. Ser. No. 92
Paper presented at Int. Conf. Optical Radiometry, NPL, London, 12–13 April 1988

Absolute spectral responsivity 0.2 to 2.5 μm

J.L. Gardner
CSIRO Division of Applied Physics
Lindfield Australia 2070

ABSTRACT: Knowledge of the absolute responsivity of detectors of optical radiation in the photometric and adjacent spectral regions is important in diverse fields, ranging from phototherapy to optical fibre communications. This paper reviews techniques that provide for calibration of absolute spectral responsivity, transfer of an absolute calibration to any wavelength in the nominated range, and the means to maintain and disseminate a scale of spectral responsivity. The power levels considered are those typical of traditional photometry and radiometry, namely nanowatts to milliwatts.

1. INTRODUCTION

A knowledge of the response of a detector to a specific quantity of optical radiation is important in various fields. Traditional photometry is concerned with visible light at power levels typically in the range from nanowatts to a few milliwatts. In this paper, we are interested in using the techniques of photometry over a wider wavelength range, from the air-ultraviolet to the near infrared spectral regions, at similar power levels to those encountered in photometry. We will not consider techniques such as calorimetry that are used to measure high power laser light, either pulsed or CW. Measurements within this restricted range of interest are applicable in a number of fields, including the calibration of UV irradiance levels for phototherapy and crack detection using fluorescence, insolation, pyrometry and optical fibre power transmission.

In theory, detector responsivity can be determined from a standard source, such as a black-body at known temperature. The spectral irradiance of such a source can be calculated, provided the geometry is known. Relative spectral responsivity can be determined in principle purely from the temperature. However, small errors in temperature can lead to large errors in irradiance (several percent at ultraviolet wavelengths). To determine spectral responsivity from a standard source, the wavelength selection device must also be fully characterized. A grating monochromator, for example, will typically produce a non-uniform, partially polarized beam. Because of these complexities accurate determination of spectral responsivity is achieved only with detector-based methods.

There are three distinct tasks in the provision of a scale of spectral responsivity over the wavelength region being considered. These are the absolute determination of responsivity, the extension of absolute values to the whole wavelength range, and the transfer of the scale to practical working detectors.

2. ABSOLUTE RESPONSIVITY

The responsivity of a detector is the quotient of response (volts, amperes) for a given radiometric power (Watts). Absolute determination of spectral responsivity has been performed for many years using electrical substitution radiometers in which radiometric heating and electrical heating are alternately applied to a bolometer, ie a device responding to a change in temperature, and the radiometric power is calculated from the easily-measured electrical power. In more recent times, quantum detectors have been used (self-calibrated) to determine the radiometric power from a current measurement. Both these techniques will be discussed in detail.

2.1 Electrical substitution radiometers

A typical electrical substitution radiometer, that used at CSIRO as the primary Australian standard of radiometric power (Blevin and Brown, 1971), is shown in Figure 1. Based on a device described by Gillham (1962), it consists of a thermopile of 28 copper-constantan thermocouples attached to a blackened copper target disk. The copper disk may also be heated by an evaporated gold electrode, insulated from the disk by a thin mica washer. The device is used in a null mode, with the electrical heating adjusted so that alternate electrical and radiant heating cycles produce the same temperature rise as measured by the thermopile. The electrical heating is measured by current/potential leads attached to the heater. Lead heating, determined with a six-terminal configuration, typically accounts for 0.3% of the total heating. The main correction to be applied is for radiation reflected from the black surface. This may be of the order of 7-8% for black paints, or about 0.25% for a gold-black surface. A number of suitable black surfaces have been described in the literature (Betts *et al*, 1985). The loss by reflection can be reduced to negligible amounts by installing a gold-coated hemispherical mirror in front of the detector to return the reflected light to the receiver. The remaining limit to accuracy of these flat-receiver radiometers is determined by the non-equivalence of radiant and electrical heating. The radiant flux is absorbed at varying depths near the surface of the blackened surface. The heat produced must diffuse through the black to the heating element and hence to the thermopile. If the thermal impedance of the black is not negligible, greater losses to the surroundings occur for the radiant heating than for the electrical heating. The thermal impedance of a carbon black is such that the radiant heating will be typically undervalued by 0.4%. This figure reduces to the order of 0.04% using gold-black. The effect can be reduced to negligible proportions by operating the gold-black coated radiometer in vacuum, thus increasing the value of the thermal impedance to the surroundings. The blackened disk radiometers used at CSIRO typically have a NEP of 3nW, so that power levels above $3\mu W$ can be measured with an uncertainty of 0.1%, but a full measurement cycle may occupy 30 min, and the source must be stable over this period.

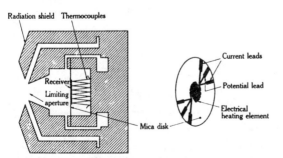

Figure 1. Electrical substitution radiometer used at CSIRO.

More modern developments of flat-plate radiometers are those described by Boivin and McNeely (1986), using lithographic techniques to produce the thermopile, and the electrically-calibrated pyroelectric radiometer (Geist and Blevin, 1973), in which the temperature transducer is not a thermopile but a pyroelectric material.

Cavity radiometers are a further extension of the flat-plate radiometer. Typical examples are those developed for broad-band solar irradiation for the World Meteorological Organisation (Brusa amd Frolich, 1986), and the Quinn-Martin cryogenic cavity radiometer (1985), which has an accuracy of .01%. Electrical substitution radiometers have the advantage of near-constant responsivity over a wide wavelength range, but they are generally slow to respond and are relatively insensitive.

2.2 Silicon Absolute Radiometry

The rare-gas ion chamber (Samson, 1967) has been used as an absolute detector of radiometric power at vacuum-ultraviolet wavelengths (<102nm, the onset of ionization in xenon), based on the fact that each absorbed photon produces one electron, at least at photon energies less than the onset of double-ionization. This technique was extended to solid-state silicon photodetectors at the US National Bureau of Standards (Geist *et al*, 1980). A typical silicon photodiode consists of an oxide region then two doped regions of p-type and n-type silicon forming a junction at the interface. When light is incident on the structure, a fraction is reflected and the remainder is absorbed in the silicon at varying depths depending on the wavelength of the light. Short wavelength light (typically < 500nm) is absorbed strongly near the surface, within the junction region. Longer wavelength light penetrates further into the silicon, away from the junction region.

The light reflected from the surface represents an external loss. Each photon of light absorbed in the silicon produces one electron-hole pair (or possibly more at short wavelengths due to multiple ionisation). If all these electrons are collected, a simple current measurement tells us the number of carriers (electrons or holes) formed per second, hence the number of photons absorbed per second. Each photon has fixed energy determined by the wavelength of the light and so we can calculate the incident flow of energy, that is, the incident radiometric power.
Two major loss mechanisms prevent the collection of all the carriers at the photodiode junction. Positive charge trapped within the oxide layer can modify the field at the surface so that the electron-hole pairs recombine before collection. Similarly, carriers formed some distance from the diode junction within the bulk material are subject to only a weak electric field and recombination may take place before collection as these carriers drift to the junction region.

Recombination in the bulk region is overcome by reverse-biasing the photodiode to form a depletion region of a high field strength to rapidly sweep the carriers to the junction for collection. Some tens of volts of reverse bias may be required at the longer wavelengths, such as the 633nm He-Ne laser line. The level of bias required is easily checked by the photocurrent showing a saturation plateau as the bias is increased. Front surface recombination at short wavelengths was a problem with the opn photodiodes to which the silicon "self-calibration" technique was first applied, usually the EG&G UV444B type. A front surface bias was applied by corona spraying of charge onto the surface or a weakly-conducting drop of water was placed onto the surface as a transparent electrode (Geist *et al*,

1982). Both of these techniques were shown to alter the surface such that the surface loss was not reproducible and was in fact changed by the oxide biasing process (Verdebout, 1984; Key *et al*, 1985).

Silicon self-calibration has been successful using onp inversion layer photodiodes, such as the UDT UV100 type. These diodes have a shallow junction and the positive surface charge helps sweep the carriers to the junction. The only biasing technique required is the reverse bias to determine recombination losses in the bulk material. The shallow junction means that long wavelength light may require large amounts of reverse bias, greater than the breakdown voltage, to obtain saturation. We have seen the onset of breakdown at bias levels as low as 17V in some of these diodes. These diodes also tend to be non-linear at high levels of irradiance and can exhibit an induced forward bias, so that they should always be operated with a small amount (about 1V) of reverse bias at moderate flux levels.

While the silicon self-calibration technique can be applied using incoherent sources and filters or monochromators (Hughes, 1982), it is most accurate with monochromatic sources such as lasers where the wavelength is accurately known. Use of the laser introduces the possibility of coherent reflections within the window usually present to seal the detector package, and between the window and the silicon surface. Variations in detected power at the level of a few percent may be seen with windowed detectors. However, careful application of the technique using a windowless detector can achieve accurate power measurements at the uncertainty level of 0.1% using argon- and krypton- ion lasers and helium neon lasers. The laser beam must be expanded to a few mm in diameter to reduce the level of irradiance, possibly attenuated to the level of 1-10 mW and actively stabilised in intensity using a servo-control loop and acousto-optic or electro-optic modulator. Light diffracted from apertures may need to be collected and imaged onto the detector with a lens. The beam will be Gaussian in profile, and if the detector is non-uniform this can lead to errors. Some of these problems may be avoided by imaging an aperture in an integrating sphere, illuminated by the laser, onto the detector.

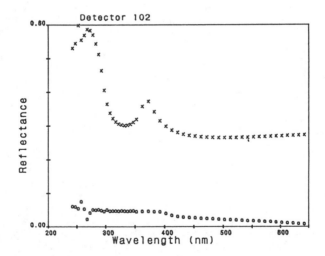

Figure 2.   Total reflectance (upper curve) and ratio of diffuse to total reflectance (lower curve, x 10) for a silicon photodiode.

The reflection loss from the front surface must be measured to an uncertainty approaching 0.2% at 500nm to obtain the incident power with an error of not more than 0.1%. Special reflectometers may be required to achieve this accuracy (Shaw and Blevin, 1964). Particularly at the shorter wavelengths, the diffuse components of reflectance may not be negligible; this may be measured relative to the total reflectance using an integrating sphere. Typical values of the ratio diffuse to total reflectance and of the total reflectance for a windowless silicon photodiode are shown in Figure 2. Packages of three specially selected diodes mounted to inter-reflect light from one to the other, so that the external reflectance loss is negligible (Zalewski and Duda, 1983), are also available for use as an absolute detector provided the incident wavelength is known.

Figure 3 shows the quantum yield of an unbiased inversion layer photodiode and illustrates the range of applicability of the self-calibration technique. At long wavelengths, the yield drops below unity because of bulk recombination. The rise at short wavelengths is due to multiple ionisation within the silicon. It is hoped that an improved knowledge of the physics of the silicon solid state will lead to accurate prediction of the quantum yield in the short wavelength region so that the technique can be extended to shorter wavelengths.

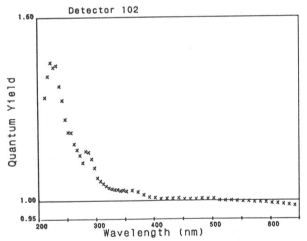

Figure 3. Quantum yield of an unbiased inversion layer photodiode.

3.   RELATIVE SPECTRAL RESPONSIVITY

Absolute spectral responsivity using either of the two above techniques is generally performed at laser wavelengths where the high power levels are useful for the insensitive substitution radiometers and the precise value of the wavelength is important for silicon radiometry. Either technique can be used with a lamp and monochromator or filters. The zero-external reflectance device is particularly useful with narrow-band filters or a monochromator, provided the beam dimensions and divergence are matched to the device and the wavelength is accurately calibrated. The use of silicon radiometry is restricted to the wavelength range where the quantum yield, with reverse biasing, is unity. In practice this corresponds to approximately 400-800nm.

The technique used at CSIRO to transfer the absolute spectral responsivity values from broadband or high power level sources to any wavelength in the

200-2500nm wavelength region is to use a gold-black coated bolometer (Blevin and Brown, 1965) as a device whose spectral responsivity is almost independent of wavelength. These bolometers do not have electrical heaters and their absolute response must be calibrated by other means. They have a small thermal mass and hence faster response time than the electrical substitution radiometers and have a much greater sensitivity.

The bolometers currently in use are deposited on an aluminium oxide substrate approx. 70 nm thick, formed by anodising and etching aluminium foil. The thin substrate is attached to a glass block with a central hole, over which a gold resistor (approx 1 ohm) is evaporated. A coating of gold black acts as the receiver. The unit is sealed to prevent cooling by air currents. The bolometer is used in a Wheatstone bridge circuit, with the current in the bolometer arm chosen to elevate the temperature of the blackened receiver to improve the sensitivity. Radiant heating produces a resistance change and the imbalance in the bridge is measured with a phase-sensitive detector, the source being chopped at a frequency of 15Hz. A copper finger behind the substrate acts as a radiation sink to improve the uniformity of response over the detector area. The bolometers have a sensitive area of 4x2mm and their NEP is approx. $0.6nW/\sqrt{Hz}$, which is within a factor of 2 of the combined Johnson noise for a 1 ohm resistor and current noise from the bridge current. A resistor of comparable value to that of the bolometer element produced a measured noise level Of $0.45nV/\sqrt{Hz}$ when substituted in the bridge.

The responsivity of the bolometer is approximately constant with wavelength, but significant corrections are made both for the window transmittance and for the radiation reflected from and transmitted by the bolometer element (both these factors are measured with an integrating sphere, since both are non-regular, ie diffuse). Typical values are plotted in Figure 4 as correction factors for the bolometer responsivity. Note that it is only the relative correction factor that is important.

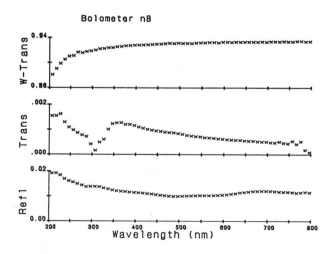

Figure 4   Reflectance, transmittance and window transmittance of a gold-black bolometer.

These bolometers require transformer coupling to match the input impedance of the next amplifier stage and the absolute gain of the coupling may be difficult to determine, and will be frequency dependent. Chopping of the beam will in general introduce a variable duty-cycle between different measurement runs. Thus the bolometer is used purely for measurements of relative spectral responsivity. Determination of the absolute responsivity of a detector is obtained by direct comparison to one of the absolute radiometers at given wavelengths rather than direct calibration of the absolute bolometer responsivity.

4. PRACTICAL MEASUREMENTS OF SPECTRAL RESPONSIVITY

The generation of a scale of spectral responsivity over a wide wavelength range will usually rely on thermal detectors. Such detectors typically have a slow response time and are relatively insensitive compared to quantum detectors such as silicon or germanium photodiodes. The gold-black bolometers described above have a small area compared to many detectors and have relatively poor uniformity. The size, shape and distribution of the optical beam used for their characterization should be closely reproduced whenever they are used to determine optical power. Hence the spectral responsivity scale is transferred to working standards, possibly using several different types of detectors to cover the entire spectral region. These working standards will have been chosen for their uniformity, linearity and stability.

The optical system used to compare detectors at CSIRO consists of a double monochromator with prism predisperser and a spherical mirror forming an image of the exit slit near two apertures in tandem. These apertures, both filled by the beam, are separately imaged onto the detectors being compared, placed symmetrically about the axis of an imaging spherical mirror which is rotated to sequentially irradiate the two detectors. Typical flux levels available in a 4nm bandwidth at the detector using a tungsten-halogen lamp are 100nW at 350nm and 30µW at 1500nm. An argon miniarc lamp is preferred for wavelengths less than 450nm, with a flux of 1µW near 220nm. Use of different apertures in tandem allows different detector sizes to be uniformly illuminated in a well-defined beam size. Where a well-defined rectangular image is required, separate horizontal and vertical slits near the monochromator exit may be imaged in the detector plane, taking advantage of the astigmatic imaging of the spherical mirrors.

Different types of detectors are used as practical working standards at CSIRO. From 200-450nm, UDT UV100 inversion layer silicon photodiodes are used with no reverse bias. Penetration of light beyond the junction region is negligible at these wavelengths; zero bias operation gives better noise performance, important since flux levels are generally low in this spectral range. These detectors are also used from 450-800nm with 4V of reverse bias to overcome induced forward bias due to the high sheet resistance. From 800-1000µm, EG&G UV444B or Hamamatsu 1337-1010BQ silicon photodiodes are used unbiased. From 1000-1800nm, Judson J16-P1 germanium photodiodes are used. Above 1800nm, the bolometers are used directly. We have found the UDT UV100 photodiodes to have a stable responsivity, within 1%, over a period of years, but other photodiodes demonstrate the aging described by Stock and Heine (1985), so that periodic re-calibration against the gold-black bolometer is required.

Absolute calibration of spectral responsivity also requires care to be taken in the choice of amplifiers. The gain of the amplifiers needs to be calibrated, and care must be taken with high-capacitance detectors if chopped radiation is used. Germanium photodiodes, particularly the low impedance, large area types, require special consideration to ensure that the input impedance of the current-to-voltage amplifier is sufficiently low to avoid losses (Stock and Mostl, 1982). These gain effects will in general be independent of wavelength and not influence the measurement of relative spectral responsivity, but become important in the determination of absolute values. Temperature effects in germanium photodiodes may be wavelength dependent.

5. CONCLUSION

As in other spectral regions, thermal detectors are still important in establishing a scale of absolute spectral reponsivity over the 0.2 to 2.5μm wavelength range. Silicon radiometry has progressed to the point where it may usefully provide absolute values of spectral responsivity, at least to the 0.1% level. Silicon photodiodes and other quantum detectors are important transfer standards for everyday use, but the long-term stability of currently available devices is such that periodic re-calibration against primary standards is still required. Future detector developments are likely to see improvements in stability, and the self-calibration technique extended to other materials than silicon as the understanding of device physics improves.

ACKNOWLEDGEMENT

The support of F. Wilkinson and W. Brown in the preparation of this paper is appreciated.

REFERENCES

Betts D B, Clarke F J J, Cox L J and Larkin J A 1985 *J.Phys E:Sci.Instrum.* **18**,689
Blevin W R and Brown W J 1965 *J. Sci. Instrum.* **42** 19
Blevin W R and Brown W J 1971 *Metrologia* **7** 15
Boivin L P and McNeely F T 1986 *Appl. Optics* **25** 554
Brusa R W and Frolich C 1986 *Appl. Optics* **25** 4173
Geist J and Blevin W R 1973 *Appl. Optics* **12** 2532
Geist J, Schaefer A R and Zalewski E F 1980 *Appl. Optics* **19** 3795
Geist J, Farmer A J D, Martin P J, Wilkinson F J and Collocott S J 1982 *Appl. Optics* **21** 1130
Gillham E J 1962 *Proc. Roy. Soc.(London)* **A269** 249
Hughes C G 1982 *Appl. Optics* **21** 2129
Key P J, Fox N P and Rastello M L 1985 *Metrologia* **21** 81
Quinn T J and Martin J E 1985 *Phil. Trans. Roy. Soc.(London)* **A316** 85
Samson J A R 1967 *Techniques of Vacuum Ultraviolet Spectroscopy*, (New York: Wiley) ch. 8
Shaw J E and Blevin W R 1964 *J. Opt. Soc. Am.* **54** 334
Stock K and Mostl K 1982 *Proc. 10th IMEKO Conf. Photon Detectors*, (Berlin) pp 40-46
Stock K and Heine R 1985 *Optik*, **71** 137
Verdebout J 1984 *Appl. Optics* **23** 4339
Zalewski E F and Duda C R 1983 *Appl. Optics* **22** 2867

Inst. Phys. Conf. Ser. No. 92
Paper presented at Int. Conf. Optical Radiometry, NPL, London, 12–13 April 1988

Space based solar irradiance measurements

Richard C Willson

Jet Propulsion Laboratory
Calif Inst of Technology
Pasadena, CA, 91109

Advances in electrically self-calibrated cavity pyrheliometry in the 1960's and 1970's, together with the accessibility of space flight observation platforms, provided the opportunity to begin space-based solar total irradiance monitoring with at least an order of magnitude better precision that earlier efforts from terrestrial, balloon and aircraft platforms. Discoveries from two data sets spanning the late 1978 to mid 1987 time period of solar cycle 21 have provided substantive new information on solar total irradiance variability on timescales ranging from minutes to the duration of the record.

The Nimbus7 Earth Radiation Budget (ERB) experiment and the first Active Cavity Radiometer Irradiance Monitor (ACRIM I) experiment on the Solar Maximum Mission (SMM) satellite have provided a nearly continuous record of solar total-irradiance variations since late 1978. Both have detected solar irradiance variations on solar active region and solar cycle timescales. The SMM/ACRIM I experiment has detected variations due to solar global oscillation modes. The ACRIM I and ERB records of long term variations revealed a downward irradiance trend during the declining phase of Solar Cycle 21.

The ACRIM I results exhibited a flat period between mid 1985 and mid 1987, followed by an upturn in late 1987 that suggests a direct correlation of luminosity and solar active region population. If the correlation is demonstrated during the upcoming increase of solar activity into solar cycle 22, it would support the relationship between the "Little Ice Age" climate and the "Maunder Minimum" of solar activity.

Two distinct regimes of correlation between irradiance and solar activity indices have been discovered from the ACRIM I results, corresponding to periods of maximum and minimum solar activity. This is a clear indication that an additional irradiance component with a solar cycle timescale is required to explain the long term irradiance variation.

Inst. Phys. Conf. Ser. No. 92
Paper presented at Int. Conf. Optical Radiometry, NPL, London, 12–13 April 1988

Ground based solar radiometry

Claus Fröhlich

Physikalisch-Meteorologisches Observatorium Davos, World Radiation Center
CH-7260 Davos-Dorf

ABSTRACT: The accuracy of ground based solar radiometry depends strongly on the fact that the instruments have to operate under ambient pressure conditions. The heat losses through the air make the temperature distribution within the cavity change for radiative and electrical heating. Hence producing a non-equivalence of the substitution and an important contribution to the uncertainty budget. A thermal finite-element numerical model of the cavity of PMO6-type radiometers has been used to get new insight in the physical basis of this effect and its influence on the performance.
Ground based solar radiometry is widely used in meteorology and climatology where comparability is more important than absolute accuracy. Thus, for the standardization the so-called World Radiometric Reference (WRR) has been introduced. The present status, the maintenance and the absolute accuracy of the WRR is discussed in the context of modern radiometry.

1. INTRODUCTION

Ground based solar radiometry was initialized and is widely used by meteorologists and climatologists for calibration of and in the radiation networks. Its use started by the end of last century with the availability of pyrheliometers, as the instruments for solar radiometry are called. The main objective was the understanding of the details of the Earth's energy balance at the surface - global, regional and local - for which the solar radiation plays obviously an important role. From the beginning it was realized that the radiometric accuracy was a limiting factor, even today with the improvements available this is still true. To go round this problem the user community made several attemps to standardize instruments and to define "scales" with more or less success. Already the attempt to define a "scale", such as the Smithonian Scale 1913 for the US community, the Ångström Scale in Europe or the International Pyheliometric Scale (IPS) 1956 and since 1981 the World Radiometric Reference (WRR), may seem to the metrologists unusual to say the least. However, the objective was to maintain comparability and not to improve the absolute accuracy in the sense of traceability to the physical unity system (SI) - in this context the term "scale" is obviously misleading and has produced a lot of confusion. The success of the method as such, however, has proven its usefullness: although the absolute accuracy of the instruments used early this century was at most 1 to 2 per cent, the comparability of the measurements made at different times and places has been maintained globally and over several decades to better than 0.5 per cent, even over changes of the "scales" in use.

Today the basic objective of solar radiometry for radiation climatology at the Earth's surface is still the same. Technical applications of solar energy, however, have expanded the user community and the demands on accuracy. Even if a well maintained reference like the WRR is used as a standard the degradation of radiometers with time ask for explanation. No improvement can be expected if the physical mechanisms governing these effects are not really understood. The only ultimate proof of the correctness of the understanding - although not sufficient - is the realization of a highly accurate radiometer and the proof of its traceability to SI with the predicted accuracy. Thus, research in the field of ambient pressure radiometry is still needed for improvements. Although much higher accuracies can be achieved today by cryogenic radiometers, these instruments cannot be used for ground based solar radiometry. One of the most important and still badly understood effect in ambient pressure radiometry is the so-called non-equivalence, an effect which is inherently tied to the air in the detector. Thus, a thorough physical understanding of this effect is needed to improve this type of radiometry.

2. GROUND BASED RADIOMETERS

The absolute radiometers for ground based solar measurements have thermal detectors using an electrically calibrated heat flux transducer. The radiation is absorbed in a cavity which ensures a high absorptivity over the spectral range of interest for solar radiometry. The different types in use today have different cavity geometry and different heat flux transducers; the general behaviour, however, is the same. In the following the PMO6 radiometer will be used as an example, because it is well-known to the author and its performance and characterization has been published (Brusa and Fröhlich, 1986). Its heat flux transducer consists of a thermal impedance and of resistance thermometers to sense the temperature difference across it (cf.Figure 1). Heat developed in the cavity is conducted to the heat sink of the instrument and the resulting temperature difference across the thermal impedance is sensed. The sensitivity of the heat flux transducer is calibrated by shading the cavity and measuring the temperature difference while dissipating a known amount of electrical power in a heater element which is mounted inside the cavity. During practical operation of the instrument, an electronic circuit maintains the temperature signal constant by accordingly controlling the power fed to the cavity heater - independent of the mode, that is whether the cavity is shaded or irradiated. In an ideal instrument the substituted radiative power would be equal to the difference in electrical power as measured during the shaded and irradiated periods respectively divided by the area of the precision aperture. However, there are many deviations from this ideal behaviour and the one over the area term will have to be replaced by a more elaborate expression accounting for these effects. The description and determination of these deviations can be found in Brusa and Fröhlich (1986).

3. NON-EQUIVALENCE EFFECT IN PMO6 RADIOMETERS

This effect is due to non-equivalent heat losses of the cavity assembly during shaded and irradiated phases, because the temperature distribution is slightly different during shaded and irradiated sequences. This causes different losses during the two modes and the substitution is no longer perfect. As the losses are mainly determined by the heat transfer through the air the non-equivalence can be determined by measuring the difference

of the sensitivity in air and in vacuum, if the radiative heat transfer is
small enough and can be disregarded.

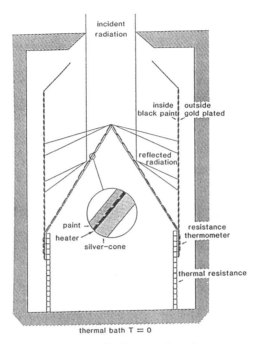

Figure 1: Schematic drawing of the cavity detector of the PMO6-type
radiometer with the subdivision into the elements used for the finite-
element calculations.

A thermal model has been generated for the cavity of the PMO6 radiometer in
order to calculate the temperature distribution for different operational
conditions with a finite-element programme. In Figure 1 the cavity is shown
schematically, subdivded in the elements used to model the thermal behaviour. The cavity is made of electro-deposited silver and is gold-plated on
their outside and coated with a thin layer of specularly reflecting black
paint on the inside. It is soldered onto the thermal resistor made from
stainless steel. The thermal resistance is in turn soldered to the copper
heat sink of the instrument. The heater element is a flexible printed circuit of 5 μm constantan foil supported by a 20 μm KAPTON foil. The thermal
model uses the thermal conductivities and geometric dimensions of the
materials used (silver, stainless steel, KAPTON, black paint) and assumes
linearized heat transfer by IR radiation from the surfaces inside of the
cavity through the hole in front to the heatsink temperature (T=0) using
for each surface element the appropriate view factor. The outside of the
cavity is coupled to the surroundings (T=0) through a surface heat transfer
coefficient α. For vacuum conditions α corresponds to the radiative transfer and amounts to 0.25 $Wm^{-2}K^{-1}$ for a gold-plated surface. For natural (not
forced) convection in air at ambient pressure α is usually taken as between
4 and 10 $Wm^{-2}K^{-1}$. The performance of the cavity is investigated by applying
heat by radiation on top of the black paint on the inverted cone with part
being reflected on the cylindrical part, heat in the electrical heater
between the black paint and the cone, and by sensing the temperture at the

place of the thermometer (mean of the temperatures at the thermometer nodes). From the results of 4 load cases, 2 for air and 2 for vacuum, the non-equivalence is calculated. With all conditions other than power input kept constant and with the power - electrical or radiative - distributed homogeneously over the surface of the impingement of the radiation, the two cases are the following:

1: 40 mW electrical heating between the black paint and the silver cone, no radiation input

2: 20 mW electrical heating between the black paint and the silver cone, 20 mW radiation input, with the fraction (1-R-r) absorbed on top of the paint on the cone, the specularly reflected fraction R absorbed on the cylinder (triangular distribution) and the diffusively reflected fraction r distributed homogeneously over the whole cylinder.

First results of the calculation showed, that the range of α values in air yields a non-equivalence which is too small. Moreover the assumptions for R (4%) and r (0%) seemed also not realistic. Laboratory experiments were performed to determine α, R and r. For the determination of air losses a normal radiometer (PMO6-9) was used and wired in such a way, that the temperature difference over the thermal impedance could be monitored. The conductivity of the whole assembly was measured by varying the power in the cavity and recording the temperature difference. No absolute values can be determined, because the thermometers are not calibrated; but only the ratio of the conductivities in air and vacuum are needed to calibrate the model. Values at intermediate pressures have also been determined in order to allow corrections to the radiometry from stratospheric balloons. Table 1 summarizes the results of the ratio of the thermal resistance normalized to ambient pressure. The α in the model was changed until the ratio between vacuum and air for a 40 mW electrical heating was achieved. An α of 25.5 $Wm^{-2}K^{-1}$ has to be used, which is compared to the engineering values a factor of 2 to 6 higher.

Table 1: Results of the determination of the heat losses of the cavity of a PMO6-type radiometer as a function of pressure, given as ratios to the value at ambient pressure (837.3 hPa)

Pressure:	0.000	1.0	3.0	10.5	837.3	hPa
Ratio:	1.3671	1.0455	1.0158	1.0073	1.0000	

The measurements of R and r were performed with a flat sample illuminated by an circularly polarized He-Ne-laser beam with at 60° incident angle. Two type of paints were investigated, a paint of a Swiss manufacturer (W.Mäder, CH-8956 Killwangen) which has been used for e.g. PMO6-9, 10 and 11 and Chemglaze (LORD Industrial Coatings, Erie, Pa.16514-0038, U.S.A.), which is now used for the PMO6-R space radiometers. For Chemglaze a diffuse part of 10 per cent of the total reflected as published by Booker (1982) has been confirmed and a value of 12 per cent has been found for the Mäder paint. The main difference between the two paints was, that the specular cone of Chemglaze is very small (<5° full angle), whereas the one of the Mäder paint is 20° full angle with an angularly integrated specularly reflected part of 8.5 to 9 per cent for both. With R=0.09, r=0.01 and α=25.5 $Wm^{-2}K^{-1}$ a non-equivalence of 0.29 per cent resulted (Brusa and Fröhlich, 1986). This is in very good agreement with the non-equivalence of PMO6-9 of 0.31. In order to investigate the influence of the opening of the specular cone, calculations were done with an opening of 20°, but the non-equivalence did not change. Quite important, however, is the effect of the diffusively reflected fraction: with R=0.10 and r=0.00 the non-equivalence decreases from 0.29 to 0.25 per cent. Thus, degradation of the black paint, e.g. by changing its surface smoothness, will change the non-equivalence and thus the correction factor and calibration. Changes of the total amount of the reflected radiation have a roughly linear effect on the non-equivalence. It is interesting to note, that the lower non-equivalence of PMO6-10 and 11 of 0.23 and 0.18 (Brusa and Fröhlich, 1986) may partly be explained by a smaller reflectivity and/or smaller part of the diffuse part of the black paint which is manifested by a smaller reflectance of the cavity. It decreases obviously with decreasing reflectivity of the paint, but the more pronounced effect in the cavity is by a change of the diffuse fraction because of the shortness of the cavity. Both the reflectance and the non-equivalence are influenced by the reflectivity of the paint and the diffuse fraction in the same direction and indeed the smaller the reflectance of the cavities of PMO6-9, 10 and 11 with 300 to 270 and 230 ppm (Brusa and Fröhlich, 1986) the smaller is the non-equivalence with 0.31, 0.23 and 0.18 per cent respectively. If the different reflectance would be the only explanation for the different non-equivalence, a change with time of the reflectance of the cavity by 0.01 per cent could change the non-equivalence and thus the calibration by up to 0.1 per cent.

From these model calculations the conclusion is that the construction of the PMO6-type radiometers is not optimal for ambient pressure radiometry. But, they indicate also ways of how their performance in air could be improved. Other type may have better performance for this kind of applications.

4. THE WORLD RADIOMETRIC REFERENCE

Radiation measurements in meteorology have to be very homogeneous in time and space in order to yield useful information about the subtle differences in radiation climates. From the beginning of the establishment of radiation networks some concern existed on how this homogeneity may be achieved. The national metrology institutes were not in a position to help, as their techniques for the establishment of radiation standards were limited to low intensity levels and, therefore, not adapted to the measurement of solar radiation. This was the main reason for different and independent development of radiation standards in meteorology. Before 1956 two "scales" were used as references for meteorological radiation measurements: the Ångström scale and the Smithsonian scale 1913 revised. In view of the existing differencies the International Radiation Commission at its meeting 1956 at Davos established the International Pyrheliometric Scale (IPS 1956), a compromise between the Ångström and Smithsonian scales. The development of absolute radiometers for solar radiometry in the late sixties suggested that such radiometer should be used to obtain a reliable reference also for meteorological radiation measurements. This was done with the introduction of the World Radiometric Reference (WRR) by the World Meteorological Organization (WMO, 1979).

Since the first participation of absolute radiometers in the 3rd International Pyrheliometer Comparison (IPC III), 1970 at Davos many comparisons between such instruments and pyrheliometers representing "IPS 1956" were performed at the World Radiation Center Davos. The results of all the intercomparisons are summarized in Figure 2. PACRAD III, an instrument developed by Kendall and Berdahl (1970), is used as reference, mainly for historical reasons. The results of the comparisons show close agreement among the 10 different types. All lie within a range of ±0.8 per cent, and half of them lie in the range of ±0.15 per cent, centered around a value which is about 0.2 per cent higher than PACRAD. The concentration of the individual results within a narrow range is an indication that the most probable value in terms of the physical unity system (SI) will also fall within this range, and that a weighted mean of these results can be used to fix the WRR. From the results of the IPCs tracing of the WRR to the former- ly used "scales" is possible with an estonishingly high accuracy. Mandatory use of the WRR has been put into the regulations of the World Meteoro- logical Organization in 1981. Its absolute uncertainty is estimated to be ±0.3 per cent.

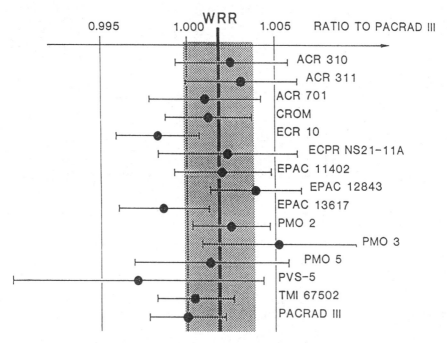

Figure 2: Summary of the results of comparisons of 15 absolute radio- meters of different design and origin and the definition of the World Radiometric Reference (WRR). The shaded area represents a ±0.2 per cent range about the WRR (Fröhlich, 1978).

The WRR is realized by a group of 6 absolute radiometers, the World Stand- ard Group: PACRAD III, PMO-2, PMO-5, CROM-2, TMI-67814, HF-14915. They all have a long enough history to know about their long-term performance and the last three IPC in 1975, 1980 and 1985 have demonstrated that the main- tenance of the WRR is within 0.1 per cent (cf.e.g. IPC VI, 1985).

The mean 7 fully characterized PMO6-type radiometers lies 0.22 per cent lower than the WRR (Brusa and Fröhlich, 1986). This is within the estimated

±0.3 per cent uncertainty of the WRR, but outside the ±0.17 per cent uncertainty of the PMO6-type radiometers. Due to the fact that none of the radiometers used to define the WRR was corrected for diffraction, the WRR is only about 0.1 per cent higher, placing it within the uncertainty of the PMO6-type radiometers.

5. ACKNOWLEDGEMENTS

The design of the cavity model and the calculations with the finite-element programme HP-FE have been performed by Stephan Tuor during his stay at PMOD/WRC, which is gratefully acknowledged.

6. REFERENCES

Booker, R.L., 1986, Specular UV reflectance measurments for cavity radiometer design, Appl.Opt. **21**, 153

Brusa, R.W. and C.Fröhlich, 1986, Absolute Radiometers (PMO6) and their experimental characterization, Appl.Opt. **25**, 4173

Fröhlich, C., 1978, World Radiometric Reference, World Meteorological Organization, Commission for Instruments and Methods of Observation, Final Report, **WMO No.490**, 108

IPC VI, 1985, Pyrheliometer Comparisons 1985, Working Report No.137, Swiss Meteorological Institute Zürich and PMOD/WRC Davos, December 1985

Kendall, J.M.Sr. and C.M.Berdahl, 1970, Two Blackbody Radiometers of High Accuracy, Appl.Opt. **9**, 1082

WMO, Technical Regulations, World Meteorological Organization, **WMO No.49**, Geneva, (1979)

Inst. Phys. Conf. Ser. No. 92
Paper presented at Int. Conf. Optical Radiometry, NPL, London, 12–13 April 1988

Current work at the National Research Council of Canada on absolute radiometer based calibrations in the infrared

L.P. Boivin
National Research Council of Canada
Physics Division, Ottawa, Canada

ABSTRACT: Hemispherical reflectors are being incorporated in some NRC absolute radiometers in order to reduce the uncertainties associated with absorber corrections. Effective absorber corrections and residual uncertainties with hemispherical reflectors are discussed. The paper also discusses the realization of a spectral irradiance scale in the infrared (700nm to 1600nm) by means of absolute radiometers used with interference filters. A new monochromator based source for absolute radiometer calibrations is described. The properties of several types of infrared detectors are discussed. The spectral characteristics of such detectors are shown to influence the magnitude of the calibration errors due to monochromator bandpass and wavelength uncertainties. A temperature-controlled radiometer is described which is very useful for the rapid determination of temperature coefficients of responsivity.

1. INTRODUCTION

Much of the current work in radiometry at NRC is directed towards expanding our facilities in the near infrared (700nm - 1800nm). Improvements to our absolute radiometers are being sought, especially for infrared use. Our present spectral irradiance scale only covers the spectral range from 300nm to 800nm. A project is under way for realizing a spectral irradiance scale in the range 700nm to 1600nm. We are also developing a new monochromator based source for calibrating directly transfer radiometers by means of absolute radiometers. In conjunction with this, we are also studying the properties of some infrared detectors in order to select those which will be suitable as transfer radiometers.

2. MODIFICATIONS TO THE NRC ABSOLUTE RADIOMETERS

The NRC absolute radiometers are of the electrical substitution type (Boivin and McNeely, 1986a), and require only one correction, for the absorber reflectance. The reflectance of the absorber (usually goldblack) is measured using a Zeiss DMC25 spectrophotometer. The uncertainty associated with this correction is less than .05% in the UV and visible. However, in the infrared, the uncertainty can be much larger (.1% to .2%) due to spectrophotometric measurement noise. In order to reduce this uncertainty, we are currently studying the use of a hemispherical reflector incorporated in the radiometer housing. This is an approach already being used in some absolute radiometers (For example, Blevin and Brown, 1971). Figure 1 gives a schematic diagram of the NRC absolute radiometer incorporating a hemispherical reflector. The latter has a radius of 35mm and is machined and optically polished from a solid piece of hard aluminium alloy (AL7075). The reflector is uncoated, and is threaded to allow accurate focussing.

The effective absorber reflectance when using a hemispherical reflector is given by

(1) $r_{eff} = r(1-R)/(1-rR)$

where r is the total diffuse reflectance of the absorber and R is the effective reflectance of the mirror surface, given by

(2) $R = K_1 K_2 R_m$

in which K_1 and K_2 are correction factors for the loss through the entrance aperture (subtending 12.8° at the receiver) and incomplete collection at large angles. R_m is the specular reflectance of the mirror, assumed to be the same as that of a flat sample of the same material, polished in the same manner. Equation

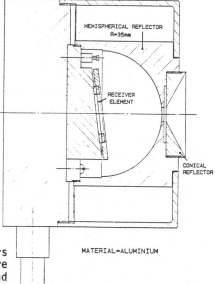

Fig 1. NRC absolute radiometer incorporating a hemispherical reflector

(2) also assumes that all the flux reflected by the mirror is received by the detector. The correction factors K_1 and K_2 are calculated from the expressions $K_1 = 1-F(6.4°)$ and $K_2 = F(80°)$, where

(3) $F(u) = \int_0^u I(x)\sin x dx / \int_0^{\pi/2} I(x)\sin x dx$

in which $I(x)$ is the angular distribution of the flux reflected by the absorber. Figure 2 shows the angular distribution of the reflected flux

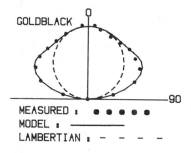

Fig 2. Angular distribution of the reflected flux for a typical gold-black absorber

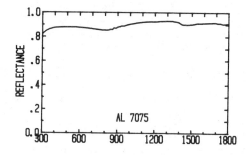

Fig 3. Spectral reflectance of a flat polished sample of type AL7075 aluminium

for a typical goldblack absorber. $I(x)$ is distinctly non-lambertian. In order to evaluate K_1 and K_2, a mathematical expression of the form

$I(x) = I_0\cos(x)(1 + Be^{-C(x-D)^2})$ was fitted to the data points. The resultant expression for the effective reflectance is given by $R = (.954\pm.015)R_m$. The uncertainty reflects the variation in the product K_1K_2 which occurs for various distributions $I(x)$ (ie for various absorbers). The variation in K_1K_2, hence the resultant uncertainty, is small. For a lambertian surface, $K_1K_2 = .958$. The values of R_m are obtained from measurements of the specular reflectance of a flat polished and uncoated sample of the mirror material, as shown in Fig. 3. R_m is about .90 throughout the visible and IR.

How does the use of a hemispherical reflector affect the uncertainty in the measurement of radiant power P? For a normal radiometer (no reflector), $\Delta P/P \sim \Delta r/(1-r) \sim \Delta r$ where Δr is the absorber reflectance uncertainty. With an hemispherical reflector, the corresponding uncertainty is given by $\Delta P/P \sim (1-R)\Delta r/(1-rR)(1-r) \sim .1\Delta r$ for a mirror having 90% effective reflectance. Thus, the use of a hemispherical mirror can greatly reduce uncertainties due to absorber measurement, ageing and thermal resistance. However, the use of a hemispherical mirror introduces an additional uncertainty component associated with the mirror itself, given by $\Delta P/P \sim -r\Delta R/(1-rR)$. For example, with a goldblack absorber ($r \sim .01$) and a mirror having 90% reflectance (with a ±2.5% uncertainty), this uncertainty component is ±.03%. At present, we have made and tested only one absolute radiometer incorporating a hemispherical reflector and using a paint absorber having a reflectance of 4.5%. Tests have shown residual errors of the order of .2% to .3%, which are attributed to a poor reflector surface quality. Work is proceeding in order to improve the quality of the optical polishing.

3. REALIZATION OF A SPECTRAL IRRADIANCE SCALE

The current spectral irradiance scale at NRC only covers the spectral range from 300nm to 800nm. A project is under way for realizing a spectral irradiance scale from 700nm to 1600nm. Absolute radiometers will be used directly with a series of interference filters in order to determine the spectral irradiance of the test lamps at a number of discrete wavelengths. Values at intermediate wavelengths will be obtained by interpolation. Figure 4 gives a schematic diagram of the apparatus.

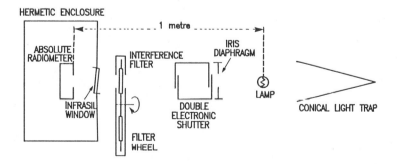

Fig. 4. Apparatus used for the realization of a spectral irradiance scale (700nm - 1600nm)

The absolute radiometer is enclosed in an hermetically sealed housing having an infrasil window which is tilted at 5° with respect to the axis in order to avoid inter-reflections between it and the interference filters. These are mounted in a large motor-driven filter wheel. The temperature of the filters is not controlled, but the temperature of the filter wheel is monitored; during the measurements, the filter temperature is 23C±3C. A double electronic shutter is used to minimize heating effects. A variable iris controls the field of view. The lamps used will be either 1000W FEL type or 750W NPL/GEC type tungsten halogen lamps. A large conical light trap is located behind the lamp to avoid back reflections. The lamp-radiometer distance is accurately set at 1 metre by means of a precision gauge bar.

For each filter, the spectral irradiance of the lamp at the effective filter wavelength λ_m is given by

$$(4) \qquad E(\lambda_m) = (W/A)K_aK_dK_wK_f \Big/ \lambda_m^5 (e^{C2/\lambda_m T}-1)\int_{\lambda_1}^{\lambda_2}t(\lambda)d\lambda \Big/ \lambda^5 (e^{C2/\lambda T}-1)$$

in which W/A is the total irradiance measured through the filter; K_a, K_d, K_w and K_f are correction factors for the absolute radiometer absorber reflectance, for diffraction effects, for the infrasil window transmittance, and for the thicknesses of the filter and window; $t(\lambda)$ is the measured spectral transmittance of the filter; and T is the colour temperature of the lamp. The basic assumptions associated with equation (4) are: (a) that the effective wavelength λ_m for each filter is given by the mean wavelength, i.e.,

$$(5) \qquad \lambda_m = \int_{\lambda_1}^{\lambda_2}\lambda t(\lambda)d\lambda / \int_{\lambda_1}^{\lambda_2}t(\lambda)d\lambda$$

(b) that for each filter, the corresponding spectral distribution of the lamp can be represented by a blackbody distribution of the form

$$(6) \qquad E(\lambda) = E(\lambda_m)\lambda_m^5(e^{C2/\lambda_m T}-1)/\lambda^5(e^{C2/\lambda T}-1)$$

(c) and that the spectral distribution of the lamp has no fine structure, no absorption and no emission lines.

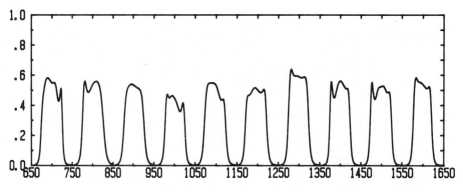

Fig. 5. Spectral transmittance curves of the interference filters

The spectral transmittances of the 10 interference filters are shown in Fig. 5. The center wavelengths of the filters are approx. 700nm, 800nm, etc. The bandwidths and peak transmittances are 50nm and 50% on average. The outer surfaces of the filters are flat to better than 10 fringes; the uniformity in the central 20mm diameter region is ±1%; the filters have an overall diameter of 50mm. Out of band transmission was measured to be less than .03% from 400nm to 5000nm. In order to reduce errors caused by filter imperfections, the transmittances of the filters will be measured using a configuration which closely duplicates the geometry used in the actual spectral irradiance measurements.

The uncertainties associated with the realization of a spectral irradiance scale by the above method result from both theoretical assumptions and measurements. The theoretical uncertainties are due mainly to the blackbody assumption (equation (6)) and the effective wavelength assumption (equation (5)). Calculations done using blackbody or tungsten radiator models at various temperatures show that theoretical uncertainties should be less than .1%. Measurement uncertainties are from many sources. Some of the sources of uncertainty and the corresponding magnitudes of uncertainty are the following: wavelength shift errors in the filter transmittance data, due to spectroradiometer errors or temperature variation of the filters (±.1%); filter transmittance measurements (±.2%); infrasil window correction (±.15%); absolute radiometer measurement (±.2%); measurement repeatability (±.2%); interpolation uncertainty (±.2%). Combining the above uncertainties in quadrature, one obtains an overall uncertainty of approximately ±.5%. This gives an estimate of the accuracy we expect to achieve in this realization of a spectral irradiance scale in the spectral range 700nm-1600nm.

4. DETECTOR CALIBRATIONS IN THE INFRARED

We are currently developing a new monochromator based source for the direct calibration of transfer radiometers by means of absolute radiometers over a spectral range from 250nm to 1800nm. Some of the characteristics of the new facility are as follows:
Monochromator: single, off-axis Czerny-Turner, high-efficiency (f3.6), $\frac{1}{4}$ meter, having a nominal wavelength accuracy of ±.2nm and nominal stray light level of 10^{-4}.
Sources: condenser coupled 250W tungsten-halogen (800nm-1800nm) and elliptical mirror coupled 75W superquiet xenon arc (250nm-800nm) and low pressure mercury vapour (wavelength calibration).
Output optics: order sorting filters; off-axis focussing optics composed of a spherical and toroidal mirror allowing beam steering capability and good quality wavelength independent focussing of a 3mm diameter beam @ f8.
Throughput: minimum 30μW into a bandwidth of 10nm or less from 250nm to 1800 nm.

In conjunction with the above, we have also been studying the properties of some infrared detectors in order to select some which are suitable as transfer radiometers. Here we shall discuss the spectral characteristics of detectors, which influence the accuracy achievable in monochromator based calibrations, and describe a temperature-controlled radiometer which is highly suitable for the rapid determination of temperature coefficients of responsivity.

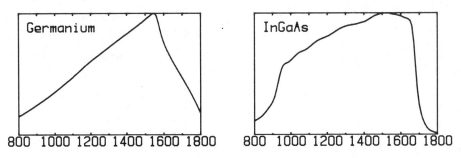

Fig. 6. Relative spectral responsivities of typical germanium and InGaAs detectors.

Figure 6 shows the relative spectral responsivities of typical germanium and InGaAs detectors. These detectors were measured using the NRC facility for routine detector calibrations (Boivin et al., 1986b). An important application at NRC of accurately calibrated transfer radiometers is to fully characterize this facility in order to improve its calibration accuracy. Silicon, germanium, and InGaAs detectors have a strong wavelength dependence. Thus, in monochromator based measurements, an important consideration is how the wavelength accuracy affects the

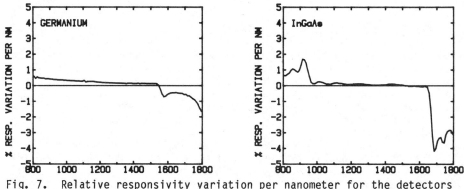

Fig. 7. Relative responsivity variation per nanometer for the detectors of Fig. 6.

responsivity calibration accuracy. This is best seen by plotting the relative responsivity variation per nanometer, as shown in Fig. 7 for the same detectors as in Fig. 6. Assuming a wavelength accuracy of ±.2nm, the maximum responsivity error would be less than .2% from 800nm to about 1750nm, for the germanium detector. For the InGaAs detector, the error would be very small (less than .1%) between 1000nm and 1600nm, but below, and especially above this range, the error would be much larger (.5% or more)

The finite bandwidth of the radiation emerging from a monochromator can also cause errors. The measured, or effective, responsivity (assuming a neutral thermopile type reference detector) is given by

(7)
$$S^*(\lambda_0) = \int_{\lambda_0-\Delta}^{\lambda_0+\Delta} S(\lambda)I(\lambda)d\lambda / \int_{\lambda_0-\Delta}^{\lambda_0+\Delta} I(\lambda)d\lambda$$

in which $S(\lambda)$ is the true responsivity, Δ is the bandwidth of the monochromator (FWHM), and $I(\lambda)$ is the spectral distribution of the monochromator output. To a good approximation, $I(\lambda)$ can be taken to be the triangular slit function (assuming equal entrance and exit slits), centered at $\lambda = \lambda_0$, and zero at $\lambda_0-\Delta$ and $\lambda_0+\Delta$. The relative bandwidth error is given by

(8)
$$e = (S^*(\lambda_0)-S(\lambda_0))/S(\lambda_0)$$

We have done calculations for typical silicon and germanium detectors, using gaussian quadrature to evaluate equation (7), in which $S(\lambda)$ was represented by accurate polynomial fits to measured $S(\lambda)$ curves. These calculations show that for a bandwidth of 10nm or less, the errors are less than .1% everywhere except near the bandgap. For a 10nm bandwidth, maximum errors of ~.12% are observed for Si($\lambda \sim 1000$nm) and of ~.14% for Ge($\lambda \sim 1550$nm). We expect that for most detectors, errors in responsivity measurements due to bandwidth effects are less important than those due to wavelength errors.

Figure 8 shows a schematic diagram and circuit diagram of a temperature-controlled radiometer head which is very useful for regulating

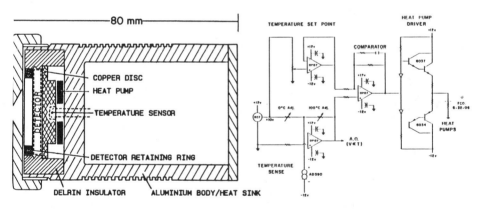

Fig. 8. Schematic diagram and circuit diagram for a temperature-controlled radiometer head.

the temperature of various types of detectors and for rapidly determining their temperature coefficients of responsivity. Two thermo-electric heat pumps are used to heat or cool a copper disc in good contact with the detector. Temperature stability is ±.1C and the radiometer can operate at any temperature between 10C and 40C. One of the most useful features of the radiometer is the speed of response (about 5 to 10 minutes for a 10C change in set point). Figure 9 shows the temperature coefficients of responsivity for a 10mm diameter germanium detector and a 3mm diameter InGaAs detector. Below 1550nm, the temperature coefficient of the

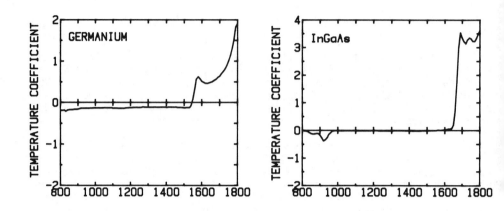

Fig. 9.   Temperature coefficients of responsivity at 23C for a 10mm germanium and a 3mm InGaAs detector.

germanium detector is nearly constant, at a value between .1 and .2%; however, above 1550nm, the coefficient increases rapidly to as much as 2% at 1800nm.   On the other hand, the temperature coefficient of InGaAs is nearly zero between 1000nm and 1600nm.   Above 1600nm, it becomes very large, approaching 4%.

Although InGaAs is in many respects a much better detector than germanium, the latter may be more useful as a transfer radiometer because it is available in larger sizes, and because its useful spectral range is larger.   For good quality germanium detectors, calibration uncertainties in monochromator-based calibrations due to wavelength errors, bandwidth effects and temperature variations can easily be kept below .1% below 1550nm; however, in the range 1550nm-1800nm, total uncertainties from the same sources can easily exceed 1%.

ACKNOWLEDGMENTS

The author wishes to thank Peter Grant, Mike Kotler and Allan Cameron for the electronic design, mechanical design and assembly of the temperature-controlled radiometer.

REFERENCES

Blevin W R and Brown W J, 1971, Metrologia 7, p.15.
Boivin L P and McNeely F T, 1986a, Appl. Opt. 25, p. 554.
Boivin L P, Budde W, Dodd C X, and Das S R, 1986b, Appl. Opt. 25, p. 2715.

Inst. Phys. Conf. Ser. No. 92
Paper presented at Int. Conf. Optical Radiometry, NPL, London, 12–13 April 1988

Radiometric determination of the gold point

N P Fox, J E Martin and D H Nettleton

National Physical Laboratory, Division of Quantum Metrology
Teddington, Middlesex TW11 0LW, UK

ABSTRACT: This paper describes the feasibility of a radiometric determination of the freezing point of gold. It also presents an assessment of the expected uncertainties of the determination and concludes that these will be approximately 80 mK.

1.   INTRODUCTION

The temperature of the freezing point of gold, 1337.58 K as defined by IPTS 68 (International Practical Temperature Scale of 1968), has come under close scrutiny in recent years. Most recent realisations of the thermodynamic temperature of the gold point have shown differences from this defined value, some as large as several hundred milli-Kelvins. However, most of these realisations are not absolute. With the imminent introduction of a new International Temperature Scale, in 1990, efforts have been renewed into making absolute realisations of this high temperature fixed point.

The spectral radiance, L_λ, of a black body at a given temperature, T, and a given wavelength, λ, can be calculated from the Planck equation:

$$L_\lambda = c_1 \pi^{-1} n^{-2} \lambda^{-5} (\exp(c_2/\lambda T) - 1)^{-1}$$

where c_1 and c_2 are the first and second radiation constants respectively and n is the refractive index of air. Hence, if the radiance can be measured absolutely for a defined wavelength interval then the temperature can be calculated absolutely. For this determination to be of value the gold point must be measured with an uncertainty of less than 100 mK, corresponding to about 0.1% for the radiance measurement.

This paper describes the feasibility of such an experiment, that is, an absolute determination of the temperature of the freezing point of gold by filter radiometry. Also included is an assessment of the uncertainties associated with the technique and some preliminary results.

2.   METHOD

The proposed experimental arrangement for the determination is shown in Fig. 1. The emitting aperture of a black-body furnace is imaged onto a radiometer with a simple lens. The uniform field of black-body radiation within the cavity of the furnace and the principle of 'constant

brightness' of a source means that there is great flexibility in the
positioning of the source, lens and radiometer when making the
measurements.

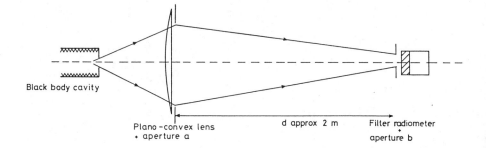

Fig. 1. Block diagram of the experimental arrangement.

The optical system can be described in two ways. Either by considering the
lens imaging the black-body cavity onto, and overfilling, the radiometer
aperture or, more revealingly, the lens imaging the aperture on the
radiometer into the cavity. The latter description shows that it is only
necessary to know the area of the aperture on the radiometer to make
radiance measurements; it is not necessary to know the emitting area of
the black-body which is hot and inaccessible. Also, by selecting
appropriate positions for the source, lens and radiometer, some
magnification of the black-body cavity is possible so that larger
apertures can be used on the radiometer. This reduces the uncertainty in
the measurement of the aperture area. Similarly, the uncertainty in
measuring the distance between the lens and the radiometer can be reduced
by using a lens with a long focal length. In this experiment the total
optical path length is about 2.5 m, the magnification can up to three
times and the radiometer aperture can be up to 6 mm diameter. A large
diameter lens was chosen to achieve a large collecting solid angle,
thereby not only increasing the accuracy of the angle measurement but also
increasing the measured signal. The limiting aperture on this lens can be
as large as 60 mm diameter.

Absolute measurements of the radiance of the black body only require the
measurement of the collecting solid angle (defined by the apertures on the
lens and radiometer and the distance between them), the area of the
aperture on the radiometer, the transmission of the lens and the spectral
response and bandpass of the radiometer.

3.   UNCERTAINTIES

The uncertainties in the measurement will be made up of contributions from
each component in the measurement system. The black-body furnace has been
described by Coates and Andrews 1978 and gives contributions to the
uncertainty from the blackness of the cavity and the stability of the
temperature at the freeze. The lens transmission measurement will
contribute to the uncertainty and diffraction effects at the aperture
edges must be considered. The measurement of the collecting solid angle
requires measurement of the distance between the two apertures, a and b in
Fig. 1, which can be achieved, using an interferometer, to better than
0.1 mm. It also requires the measurement of the aperture areas and this

together with associated uncertainties has been described by Goodman *et al* 1988. Finally there is a contribution to the uncertainty from the radiometer. The radiometer has been described previously by Fox *et al* 1986 and can be operated at two wavelengths, 676 nm and 800 nm, with a spectral bandpass set by interference filters of approximately 20 nm. It has been characterised so that its reponse can be calibrated with an uncertainty of $\pm$ 0.15% but it is expected that this uncertainty can be reduced by further investigations.

A summary of the uncertainties, at the one standard deviation level, associated with the determination are presented in the table.

	PARAMETER		VALUE	UNCERTAINTY (%)	(mK)
SOURCE	Emmisivity		0.99999	0.005	5
	Purity of Au		0.99999	0.001	1
	Temp. uniformity			0.02	20
DETECTOR	Absolute scale			0.02	20
	Spectral characterisation			<0.05	<50
	Nonlinearity			0.02	20
	Amplifier gain			0.02	20
	Spatial uniformity			0.03	30
METHOD	Distance		1800 mm	0.005	5
	Area aperture a	i	2830 mm^2		
		ii	1960 mm^2	0.01	10
		iii	1260 mm^2		
	Area aperture b	i	28 mm^2		
		ii	19 mm^2	0.01	10
		iii	12 mm^2		
	Transmission of lens		0.992	<0.05	<50
	TOTAL IN QUADRATURE			0.08	80

4.   RESULTS

The results presented are not of a gold freezing point determination but show the feasibility of such a determination.

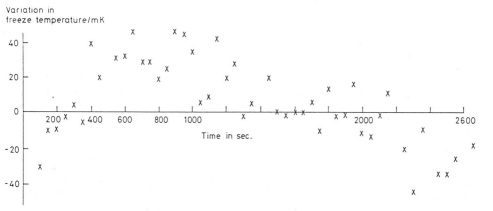

Fig. 2. The black body spectral radiance at the freezing point of gold.

Fig. 2 is a plot of the spectral radiance of the black body, measured by the 800 nm filter radiometer, as a function of time for a single freeze of the gold point furnace and shows that the freeze temperature can be identified to ± 20 mK.

Fig. 3 shows the spectral radiance of the black body plotted as a function of the distance between the lens and the radiometer, keeping the distance between the lens and the black body constant. This corresponds to moving the image of the radiometer aperture in and out of the black body cavity. The plot clearly shows the uniform black-body radiation field within the furnace cavity and confirms the insensitivity of the measurement to the relative positions of the source, lens and radiometer. The positions where the image is 'clipped' by the cavity's aperture can also be clearly seen.

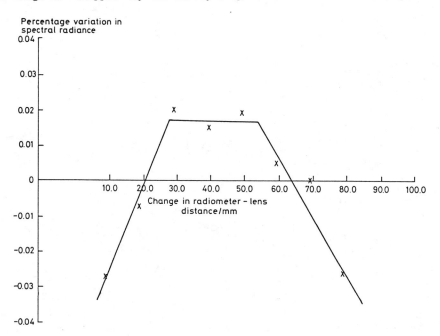

Fig. 3. The uniformity of the spectral radiance within the black body cavity.

5.   CONCLUSION

Using the optical system described above it should be possible to determine the temperature of the freezing point of gold absolutely by a radiometric measurement with an uncertainty of 80 mK.

6.   REFERENCES

Coates P B and Andrews J W 1978 *J. Phys. F: Metal Phys.*, **8** No. 2 pp 277-285
Fox N P, Key P J, Riehle F and Wende B 1986 *Appl. Opt.* **25** pp 2409-2420
Goodman T M, Martin J E, Shipp B D and Turner N P 1988 Elsewhere in these proceedings

Inst. Phys. Conf. Ser. No. 92
Paper presented at Int. Conf. Optical Radiometry, NPL, London, 12–13 April 1988

Application of absolute radiometry to the measurement of optical power in fibre optic systems

D H Nettleton

National Physical Laboratory, Division of Quantum Metrology
Teddington, Middlesex TW11 0LW, UK

ABSTRACT: Facilities have been developed at the National Physical Laboratory (NPL) for the calibration of commercial fibre optic power meters. A special transfer standard detector has been developed incorporating an integrating sphere and a fibre optic connector which has been calibrated against the NPL responsivity scale with an uncertainty of ± 0.25% for single mode fibre and ± 0.5% for multimode fibre. A near infrared relative spectral responsivity scale has also been established with an uncertainty of approximately ± 0.3%.

1. INTRODUCTION

Fibre optic telecommunication systems operate in the near infrared using the silica spectral windows at 850 nm, 1300 nm and 1550 nm over the power range milli-watts to nano-watts. Optical power is measured in these systems with commercially produced meters fitted with a suitable connector. These power meters require calibration against national standard laboratory responsivity scales. It is difficult to couple optical power to the meters from traditional responsivity calibration facilities and so special facilities and transfer standards have to be developed.

A calibration strategy was chosen to minimise the requirement for absolute responsivity calibration to a few measurement conditions. Other measurements required to fully characterise the response of a meter can be made relative to these absolute measurement points, relaxing some measurement conditions. It was decided to make absolute response measurements at only one high power level at each of the silica fibre spectral windows: 850 nm, 1300 nm and 1550 nm. Relative measurements would then be made to check the response at lower power levels and at other wavelengths.

2. ABSOLUTE RESPONSIVITY MEASUREMENT

The NPL absolute responsivity scale is established by a cryogenic radiometer described by Martin, Fox and Key (1985) with an uncertainty of less than ± 0.005% using intensity stabilised lasers. Thermopile radiometers designed by Preston (1971), and now produced commercially by Laser Instrumentation Ltd in the UK as the 14BT model, are calibrated against this visible scale using the same intensity stabilised lasers at a power level of about 1 mW and with an uncertainty of ± 0.1%. Any spectral dependence of the response of these radiometers is due to changes in reflectance with wavelength of the black absorbing disc onto which the radiation is incident. The reflectance of this disc, which is painted with black Nextel paint of the 3M Company, has been measured at NPL using spectrophotometric techniques. The measurements show that the reflectance is spectrally flat to about $\pm$ 0.1% from the visible out to 1700 nm. The visible response measured against the cryogenic radiometer will therefore be maintained in the infrared with only a slight increase in uncertainty to less than $\pm$ 0.2%.

It is not possible to use these thermopiles to calibrate fibre optic power meters directly because of the sensitivity of their response to objects in the field of view. Solid-state photodiodes are more suitable detectors for coupling to optical fibre but have a response which varies with wavelength. An automated facility has been constructed to calibrate the response of solid-state photodiodes against the 14BT thermopile. The low sensitivity of the thermopile limits measurements to power levels of 100 µW or more. This is not achievable from spectrally dispersed incandescent sources and so commercial laser diode sources were used. These are available with narrow bandwidths at each of the required wavelengths: 850 nm, 1300 nm and 1550 nm. The lasers launch upto 1 mW of radiation into a fibre optic cable. The output from the cable is then imaged onto the detectors in turn using a collimating and an imaging optic designed for use with near infrared radiation.

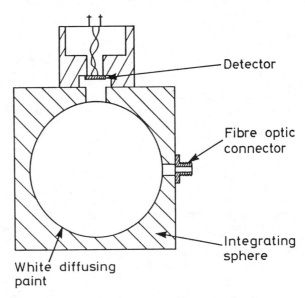

Detector

Fibre optic connector

Integrating sphere

White diffusing paint

Fig. 1. Fibre optic integrating sphere transfer standard

Early measurements indicated variations in the measured response of the photodiodes at the 1% level depending on the image position relative to the detector sensitive area. Some of this variation, if not all, could be explained by the known non-uniformity of response of germanium photodiodes across their sensitive area and the power density dependent linearity of some silicon photodiodes at near infrared wavelengths as reported by Schaefer, Zalewski and Geist (1983). To overcome this limitation the solid-state photodiodes were mounted onto integrating spheres of 50 mm diameter. This made the detector response both insensitive to input patch size and position. It was also possible to incorporate a fibre optic connector at the input port of the sphere. An FC-type connector was chosen with the central aligning components removed but in principle any connector type could be fitted and interchanged without altering the calibration. A schematic drawing of the integrating sphere transfer standard detector is shown in Figure 1.

Initially 5 mm germanium photodiodes have been used with the integrating sphere. These provide a reponse over the three spectral windows of silica fibre but, because the integrating sphere attenuates the signal by about three orders of magnitude, have a low signal to noise level. Despite this they have been calibrated with an uncertainty of ± 0.25% for single mode fibre and ± 0.5% for multimode fibre. Improvement in performance is expected by replacing the germanium photodiodes with silicon photodiodes at 850 nm and InGaAs photodiodes at 1300 nm and 1550 nm.

Commercial power meters can be calibrated for absolute response against these transfer standards by direct substitution using the same laser diode sources and connecting optical fibre cables.

3.   LINEARITY MEASUREMENT

The transfer standards described in section 2 are not suitable for measurements at power levels less than about 100 µW. These measurements are made against a cooled 5 mm diameter germanium photodiode fitted with a cap incorporating a fibre optic connector. Typical linearity data for this type of detector is shown in Figure 2 and the reduction in noise with cooling makes measurements down to about 0.1 nW possible.

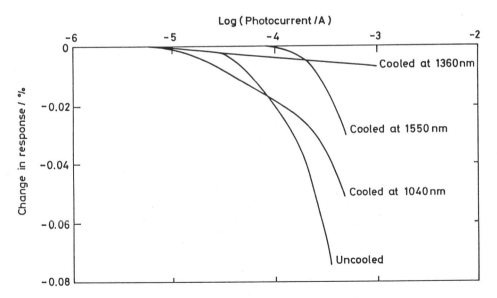

Fig. 2. Typical linearity data for germanium photodiodes

Power meters can be calibrated by direct substitution with this cooled detector using a laser diode source and connecting optic fibre cables. A variable optical attenuator is used to select the required power level. By relating all measurements to the absolute measurements at 100 µW long term stability requirements for detector response are unimportant. However the substitution method of measurement is laborious requiring many connections and disconnections of fibre optic cables. Automation has been achieved using commercially available computer controlled attenuators, measuring and recording set losses with the cooled germanium photodiode, connecting the power meter to be calibrated and resetting the attenuator and recording the response.

4.   RELATIVE SPECTRAL RESPONSIVITY

A relative spectral responsivity scale has been established from 800 nm to 1800 nm with spectrally flat pyroelectric detectors. These detectors have a 3 mm square active area coated with a platinum black behind a silica window and were manufactured by Plessey plc in the UK. The detectors have a sensitivity which is high enough to make measurements with monochromatic radiation from a spectrally dispersed tungsten ribbon lamp source.

The spectral variation of reflectance of the Plessey deposited platinum black has been measured spectrophotometrically at NPL. The transmission of the window has also been measured. By combining these two parameters the relative spectral response of the pyroelectric detector has been calculated. 5 mm diameter germanium photodiodes have been calibrated against this scale and the combined uncertainties are shown in Figure 3.

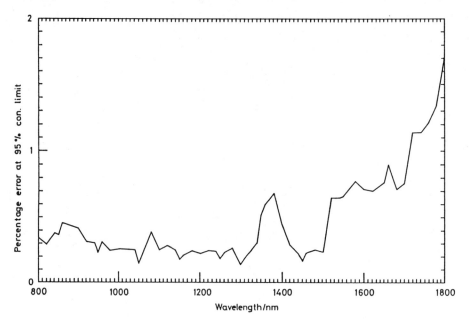

Fig. 3. Uncertainty of relative spectral responsivity scale

The increase in uncertainty around 1400 nm is due to variable atmospheric water absorption and the increase at wavelengths longer than 1500 nm is due to the increased non-uniformity in the response across the sensitive surface and to the increased temperature coefficient of germanium photodiodes.

Power meters are calibrated against these germanium photodiodes either by removing the connector cap from the meter and imaging a patch of monochromatic radiation onto the sensing element or imaging it onto the end of a fibre cable connected to the meter. Once again measurements are related to the absolute calibration, minimising errors due to different patch sizes etc.

5. ERRORS IN POWER MEASUREMENT CAUSED BY INTER-REFLECTIONS

Silicon and germanium photodiodes have high reflectivities, typically between 20% and 40%. This means that a connector cap can increase the response of a sensor head of a power meter by reflecting radiation back onto the detector. Increases in response of as much as 16% (0.64dB) have been seen. Often the photodiodes have anti-reflection coatings which make the reflectance wavelength dependent. This causes significant variation in the increase in response with wavelength. Great care must be taken, therefore, when interpreting calibrations carried out without a connector cap if one is fitted later. Also the reponse of power meters may change if caps are interchanged. For accurate power measurement calibrations should be carried out with the appropriate connector cap already fitted.

6. REFERENCES

Martin J E, Fox N P and Key P J 1985 Metrologia **21** pp 147-155
Preston J S 1971 J. Phys. E: Sci. Instrum. **4** pp 969-972
Schaefer A R, Zalewski E F and Geist J 1983 Appl. Opt. **22** pp 1232-1236

Inst. Phys. Conf. Ser. No. 92
Paper presented at Int. Conf. Optical Radiometry, NPL, London, 12–13 April 1988

99

Blocked impurity band and superlattice detectors: prospects for radiometry*

Jon Geist
National Bureau of Standards, Gaithersburg, MD 20899

ABSTRACT: Blocked Impurity Band detectors and photomultipliers, which have been described by Petroff and Stapelbroek, may be suitable for use as high-accuracy standards for low background optical radiation measurements extending from the near ultraviolet to beyond 25 μm in the infrared. The current status of their development from the point of view of standards applications is reviewed.

Superlattice technology offers new materials properties, new degrees of freedom, and new possibilities for optical radiation detectors displaying a large range of tailorability and tunability. GaAs/AlGaAs superlattices are used to illustrate new properties, HgTe/CdTe superlattices are used to illustrate new degrees of freedom, and GaAs-doping superlattices are used to illustrate tailorability and tunability.

1. INTRODUCTION

The epitaxial growth of semiconductor crystals on semiconductor substrates allows the fabrication of new types of photodetectors having properties that cannot be attained with devices based on conventional processing of a single substrate. Of particular interest to absolute radiometry are devices that have greatly improved sensitivity without an intolerable loss of the qualities that are necessary for high accuracy measurements. This paper reviews two new epitaxial technologies with respect to their impact on absolute radiometry. The first is the Blocked Impurity Band (BIB) detector technology introduced by Petroff and Stapelbroek (Szmulowicz 1987; Petroff *et al.* 1987), which is based on As-doped Si epitaxial layers. Devices of potential use to absolute radiometry have already been demonstrated in this technology.

The second new epitaxial technology of potential interest to absolute radiometry is that of the quantum-well superlattice (Doehler 1983, 1986; Capasso 1987), which is the subject of a very large, worldwide R&D effort. While this technology has not yet produced any devices of use to absolute radiometry, its potential is enormous, because it allows the fabrication of structures having properties that could not even be imagined, much less fabricated, in the conventional semiconductor materials technologies. The remainder of this paper is devoted to a review of the physics behind these two technologies, in an attempt to make the literature of these technologies

---

more accessible to absolute radiometrists.

2. BLOCKED IMPURITY BAND DETECTORS

Both BIB photodiodes (Szmulowicz 1987) and BIB photomultipliers (Petroff *et al.* 1987) have been demonstrated within the last few years. Both devices are designed to measure very low photon fluxes in low background-radiation environments. As a rough generalization, the maximum flux level should be below 10^9 photons per second, and the maximum background radiation must be below this level. This means that the BIB detector's operating temperature and radiation-environment temperature must be no greater than a few tens of kelvins. These requirements place severe limits on the potential application of these devices, but in those applications to which they are suited, their predicted radiometric performance is nearly ideal. Before describing BIB detectors in detail, we review the behavior of As-doped silicon at very low temperatures as a function of As concentration.

2.1 DONORS, IMPURITY BANDS, MOTT TRANSITIONS, AND ALL THAT

Arsenic has one more outer shell electron than Si, so when it is incorporated into the Si lattice as a substitutional impurity, four of its outer shell electrons are involved in bonds that maintain the Si lattice structure, while the fifth is loosely bound to the ionic core. The hydrogen-like, impurity-ground state is located about 54 meV below the bottom of the conduction band. If the electrons in this state absorb more than 54 meV from photons or are thermally ionized by phonons, then they enter the conduction band, and are free to move about the material, and thereby contribute to the electrical conductivity of the material. The As atoms that are incorporated into the silicon lattice substitutionally are called donor atoms because they can donate free electrons to the conduction band in this way. At room temperature, almost all of the donors are ionized by interaction with the sea of thermal phonons. As a result, silicon that is heavily doped with As is a good conductor. But at temperatures no greater than a few tens of kelvins with a radiation environment in the same temperature range, virtually none of the donors are ionized, and the material is an excellent insulator. We now restrict our considerations only to low temperatures, and consider what happens as the As concentration is varied.

At doping densities below a few times 10^{17} As atoms/cm^3, the As atoms are so far apart that there is negligible overlap between the hydrogen-like wavefunctions of neighboring As atoms. As a result the material is an excellent insulator, with a filled, narrow impurity level in the forbidden gap between the valence and conduction bands, and a hydrogen-like series of unfilled levels between the impurity-ground state and the bottom of the conduction band.

At doping densities between a few times 10^{17} As and 8×10^{18} As atoms/cm^3, the hydrogen-like wavefunction of the As atoms overlap enough that the interaction between neighboring As atoms becomes important. The ground state broadens into what is called an impurity band, and the gap between the impurity band and the bottom of the conduction band decreases with increasing doping in this range of dopant concentration. When one As atom is ionized, by absorption of a photon, for example, there is a significant probability that an unionized electron from a neighboring As atom will hop to the ionized atom as a result of the overlap between the neighboring hydrogen-like, As wavefunctions. This probability is asymmetric in the presence of an external electric field, and therefore causes the flow of an electric current. This type of current flow is called hopping mode conduction. It is important to notice that it occurs without any electrons in the conduction band, and that none of the electrons participating in the conduction process have enough energy to enter the conduction band. Also, the mobility of a hopping mode electron is many orders of magnitude lower than that of a conduction band electron, but increases rapidly with temperature.

Above 8×10^{18} As atoms/cm^3, the As atoms are so close that the interaction between neighboring atoms is large enough to cause the impurity band to merge into the conduction band. As a result, even at 0 K, some of the impurity electrons are in the conduction band where they have conduction band mobilities. Now the material behaves like a metal, rather than an impurity band conductor. The doping concentration at which this transition takes place is called the Mott transition. With this background we can now describe the generic BIB photodiode.

2.2 THE GENERIC BIB DETECTOR

An idealized BIB detector is shown in Fig. 1. The substrate of the BIB detector has a sufficient As concentration to be above the Mott transition. This is indicated in the figure by showing electrons (e) above the conduction-band edge. The first layer, called the impurity-band layer or just impurity layer, has the correct As concentration to have a well-developed impurity band. The impurity band electrons are localized on the As atoms as indicated in the figure. The second layer, called the blocking layer, is as pure as possible, to prevent the "flow" of donor ions by hopping mode conduction. The third layer has a sufficient As concentration to be above the Mott transition. The two heavily doped layers serve as electrodes, the substrate serving as the rear electrode and the third layer serving as the front electrode.

A silicon wafer cut from an ingot grown with the proper dopant concentration generally serves as the substrate, while the impurity and blocking layers are generally formed on the substrate by an epitaxial growth process. The front region electrode is generally formed by ion implantation into the blocking layer.

There will always be some contamination of the impurity layer by substitutional boron atoms and other atoms with three electrons in their outer shell. These create what are called acceptor levels near the top of the valence band. These are similar to the hydrogen-like As levels just below the bottom of the valence band except that they are negatively charged when occupied with an electron, while the As donor levels are neutral when occupied.

In the absence of a reverse bias, all of the acceptors in the impurity layer will have picked up an electron from a nearby As atom, because their energy levels are lower than those of the As donors. So all of the acceptors and an equal number, but small fraction, of the As donors will be charged, but macroscopically, the material will be electrically neutral.

When a positive voltage is applied to the front electrode, hopping mode conduction

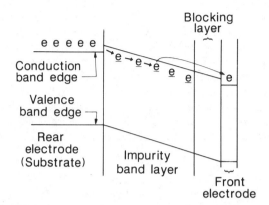

Figure 1. Idealized BIB photodiode under reverse bias. The silicon substrate was grown with sufficient As doping to be above the Mott transition, so even at 0 K all of the donor electrons (e) are in the conduction band. The substrate serves as an electrode. The impurity and blocking layers are epitaxial Si layers. The impurity layer was grown with the correct As concentration to establish an impurity band that supports hopping mode conduction. This is indicated by the short arrows connecting the electrons in the impurity band states just below the conduction band. The blocking layer was grown as pure as possible, and the front electrode was formed by ion implantation of As into the blocking layer. When a photon is absorbed by a transition of an impurity band electron into the conduction band, the electric field moves this electron to the front electrode. This process is indicated by the long arrow. Hopping mode conduction of the impurity band electrons "moves" the empty impurity band state to the conduction band where it disappears. The net effect is to move one electron across the potential difference, thereby inducing the flow of one electron in an external short circuit.

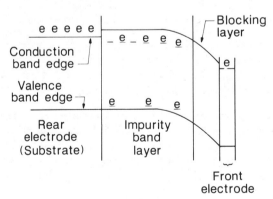

Figure 2. A more realistic model of the potential drop across a reverse biased BIB photodiode. Without bias some of the impurity band states are empty because some of the impurity band electrons have filled unwanted, but unavoidable acceptor states. The application of reverse bias moves electrons from the rear electrode to fill the empty impurity band states by hopping mode conduction. This causes the applied potential to be dropped across the resulting space charge, creating the non-uniform potential illustrated in the figure.

will allow the charged donors closest to the front electrode to pick up a fifth electron and become electrically neutral, no longer neutralizing the negative charge of nearby acceptors. As a result a space charge, whose width increases with increasing reverse bias, grows from the blocking layer into the impurity layer. This situation is illustrated in Fig. 2. It is similar to the growth of the depletion region in a reverse-biased silicon photodiode, and it is called reverse bias by analogy.

To continue the analogy further, increasing the reverse bias on a BIB detector increases the width of the space charge region that is depleted of the ionized donors that support hopping mode conduction, just as increasing the reverse bias on a silicon photodiode increases the width of the space charge region that is depleted of the free electrons and holes that support metallic conduction. At any temperature above 0 K, reverse bias causes a dark current, which increases with increasing temperature and As concentration.

Suppose an infrared photon ionizes an impurity band electron in a reverse-biased BIB photodiode, thereby moving it from the impurity band to the conduction band. Then the electric field will move that electron through the conduction band to the front electrode, and it will "move" the ionized impurity through the impurity band layer to the rear electrode by hopping mode conduction of the impurity band electrons. This means that one quantum of charge has been moved across the space charge region, which would produce the flow of one electron in an external short circuit. This situation is illustrated in Fig. 1, where arrows to the left in the figure indicate processes that follow the processes indicated by the arrows to their right.

As the width of the BIB depletion region increases, the electric field at the interface between the blocking layer and the impurity layer increases until it becomes so large that it triggers the creation of new donor-ion/conduction-band electron pairs simultaneously with the transition of photogenerated conduction band electrons to lower energy states in the conduction band. This transition is called impact ionization, and it gives rise to gain.

From the above, it is clear that the BIB device behaves as if it were a photodiode under reverse bias, with the exception that it has no space charge region in the absence of reverse bias. Also, the recombination of photogenerated carriers is very unlikely, because at low photon fluxes densities, there are very few ionized impurities available to serve as recombination sites. This means that from a functional point of view, there is no significant difference between operating a BIB device and a high-quality silicon photodiode under reverse bias. This means that the self-calibration of BIB photodiodes should be possible. In particular, the shape of the relative responsivity of a BIB photodiode as a function of reverse bias at a single wavelength should determine its absolute responsivity as a function of reverse bias at all wavelengths.

A demonstration of the feasibility of self-calibrating a BIB photodiode was recently completed in collaboration with Drs. J. Boisvert and M. Sweet of the Naval Ocean Systems Center in San Diego, California. The results were promising, and we are preparing to implement a routine self-calibration capability based on BIB photodiodes for the low-background, long-wavelength infrared. However, there are some detailed differences between self-calibrating BIB photodiodes and self-calibrating silicon photodiodes in the visible besides the difference in operating temperature and spectral region.

2.3 BIB PHOTOMULTIPLIERS

While it is essential to the accuracy of BIB photodiode self-calibration that the reverse bias across the BIB diode be small enough that the gain be negligible, gain can be used to advantage in other applications. By intentionally doping the impurity layer with acceptors in the region of the blocking layer interface, a very large field can be dropped across a very narrow region.

A gain that saturates with increasing reverse bias at a value in excess of 10^4 can be achieved in this way, and because the gain occurs only in a small region of the photodiode, there is little excess noise associated with it. This is in marked contrast to the gain in an avalanche photodiode, which generates a large amount of excess noise even at very modest gains, due to the fact that avalanche multiplication is a breakdown phenomenon, as well as to the fact that the avalanche process is spatially distributed. Thus it seems that the term BIB photomultiplier, instead of BIB avalanche photodiode, is the proper term for a BIB device built in this high-gain configuration. In fact, the BIB photomultiplier produces excellent pulse distributions for photon counting applications.

BIB photomultipliers also respond to visible photons with a near 100% quantum efficiency, and noise equivalent photon fluxes of one photon per second have been demonstrated for this spectral region. Thus the BIB photomultiplier is a device with the following characteristics. Its gain is saturable at a value in excess of 10^4, and it produces excellent pulse height distributions. The noise equivalent photocurrent is about one photon per second, and the primary quantum efficiency, which is to be multiplied by the gain to obtain the overall quantum efficiency, is unity for wavelengths from about 400 nm to 1 μm. Above 1 μm the quantum efficiency falls rapidly about 2 orders of magnitude and then starts to increase with wavelength, reaching a peak of about 30% near 24 μm. The quantum efficiency then drops to zero over the next 4 to 8 μm depending upon the As dopant concentration in the impurity layer. Except for increased quantum efficiency at all wavelengths, but particularly near 1 μm, it is hard to see how the performance of this device could be usefully improved for measurements of low photon fluxes against low radiation backgrounds.

3. QUANTUM-WELL SUPERLATTICES

Quantum-well superlattices are periodic structures made from alternate layers of two semiconductors in which the widths of the layers are small enough that quantum mechanical effects control the behavior of the device. This technology offers new materials properties, new degrees of freedom, and new possibilities for tailoring optical radiation detectors for particular applications and for building detectors with new ranges of tunability. The remainder of this section is devoted to demonstrating these points.

Figures 3 through 6 show portions of the superlattices that are considered in the following sections. Figure 3 shows the band edges as a function of depth in a compositional superlattice obtained by growing alternate layers of GaAs and $Al_xGa_{1-x}As$

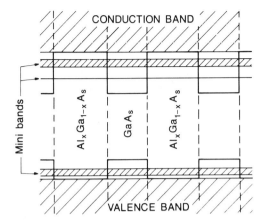

Figure 3. A portion of a GaAs/Al$_x$Ga$_{1-x}$As compositional superlattice in which neither layer is doped. GaAs and AlAs have very similar lattice constants, so good quality epitaxial layers can be grown for all values of x. The dark square-wave lines are the square-well potentials that are seen by the valence and conduction band electrons. Solution of Schroedinger's equation for these potentials produces valence and conduction bands starting near the tops of the square well barriers, and minibands within the barriers as illustrated. The mobility of the electrons in a miniband is very different in the superlattice direction than in the directions perpendicular to that direction due to the square well potential barriers.

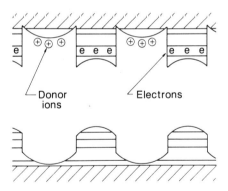

Figure 4. A portion of a GaAs/Al$_x$Ga$_{1-x}$As compositional superlattice in which the Al$_x$Ga$_{1-x}$As layer is doped. The donor electrons end up in the lowest conduction miniband, thereby creating a spatially modulated space charge to exist in the superlattice, even though it is electrically neutral as a whole. Because the wave functions of the electrons have virtually no overlap with donor ions, the mobility of the electrons in the directions perpendicular to the superlattice direction is increased over what it would be in a similarly doped crystal of GaAs. Thus, the superlattice has provided new materials properties. The minibands are shown as localized in this figure, although this would not be necessary to achieve the mobility enhancement. The miniband wave functions would be very small in the high potential region where the donor ions are located whether or not the minibands were localized.

on a GaAs substrate, with x ≈ 0.25. Figure 4 shows the band edges as a function of depth in a similar structure, where the $Al_xGa_{1-x}As$ layer has been doped with about 5×10^{16} donors per cm^3. Figure 5 shows the band edges as a function of depth in a HgTe/CdTe compositional superlattice. Notice that there is no bandgap in the HgTe layers shown in this figure. Finally, Fig. 6 shows the band edges as a function of depth in a doping superlattice obtained by growing alternate layers of donor-doped and acceptor-doped GaAs on a GaAs substrate.

The common feature of these quantum-well superlattices is a periodic array of potential wells. Some of the states having energies below the top of the potential well are not allowed. The allowed states consist of narrow bands separated by larger gaps in the energy range between the minimum and maximum of the potential well, as shown in Fig. 3. The narrow bands lying below the top of the potential well are sometimes called minibands to distinguish them from the conduction bands that lie above the top of the potential well.

In any real superlattice, the minibands will be broadened by fluctuations in the width, composition, and dopant concentration in the layers, and by applied electric fields. As long as the band broadening is less than the intrinsic width of the band, then the band will be unlocalized as shown for the bands in Fig. 3. On the other hand, if the band broadening is greater than the intrinsic width of the band, then the band will have been converted to localized states associated with individual wells, as is illustrated in Fig. 4. For more detail on this issue, see Doehler (1983).

The existence of the minibands and localized states changes the transport properties in the superlattice direction by many orders of magnitude from what they are in the

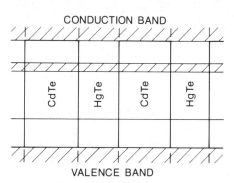

Figure 5. A portion of a HgTe/CdTe compositional superlattice. Notice that the conduction and valence bands lie on top of each other in the HgTe layer, because HgTe is a semi-metal. The long wavelength cut off of a photodetector made of this superlattice material is determined by the width of the HgTe layer, because that width determines the height of the first conduction miniband above the HgTe valence band. On the other hand, the dark current is determined by the width of the CdTe layer, because that width determines the mobility of the first conduction miniband electrons in the superlattice direction. Thus the dark current and long wavelength can be independently controlled. This is not the case with conventional $Hg_xCd_{1-x}Te$ photodetectors, so the superlattice introduces a new degree of freedom.

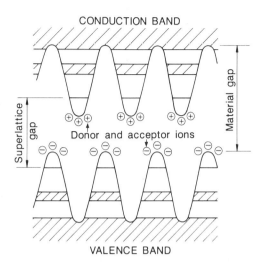

CONDUCTION BAND

Superlattice gap

Material gap

Donor and acceptor ions

VALENCE BAND

Figure 6. A doping superlattice formed by the growth of alternately donor-doped and acceptor-doped layers of GaAs. The number of donors per period is equal to the number of acceptors per period, and all of the donor electrons end up on the acceptors, This creates a spatially modulated space charge. The superlattice bandgap, which is an indirect gap in real space, is determined by the total number of donors per unit area per period. Therefore, the superlattice is highly tailorable. The concentration of carriers in the conduction and valence minibands can be controlled by electrical bias and by irradiation with photons of sufficient energy. Carriers in the minibands neutralize the space charge associated with the donor and acceptor ions. This changes the superlattice band gap accordingly, so the doping superlattice is highly tunable.

other two directions. It is possible to conceive of devices with unique properties that could not be built in homogeneous materials by taking advantage of this property of superlattices. But other, less obvious advantages are also available, as is discussed in the following sections.

3.1 NEW MATERIALS PROPERTIES

If the superlattice of Fig. 4 is grown with GaAs regions of about 10 nm width, and the $Al_x Ga_{1-x} As$ regions are doped with about 5×10^{16} donor impurities per cm^3, then, at room temperature, all of the electrons associated with the donors in the $Al_x Ga_{1-x} As$ will be thermally excited into the conduction band. From here they can make transitions to the lowest energy states in the GaAs layers. Thus all of the free electrons end up in the lowest energy states in the GaAs. There are no donor atoms in these layers to scatter the electrons during transport, so the electron mobility in the directions perpendicular to the direction of the superlattice is increased compared to that for comparable electron concentrations as shown in Fig. 7, which is based on Dingle *et al.* (1978). The increase is largest at low temperatures where scattering by donor ions is the major mechanism limiting the electron mobility.

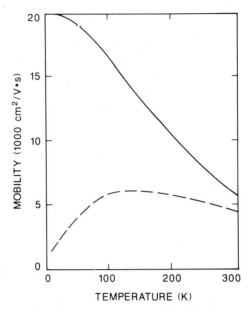

Figure 7. Mobility of electrons perpendicular to the superlattice direction for the superlattice of Fig. 4 compared with that of GaAs with a comparable electron concentration. (After Dingle *et al.* 1978)

3.2 NEW DEGREES OF FREEDOM

Notice that the HgTe layers in the superlattice shown in Fig. 5 are semimetals. That is, there is no gap between the valence and conduction bands, and the lowest conduction miniband is occupied in the absence of donor doping. The mobility in the superlattice direction of the electrons in the minibands is governed both by the height of the potential barrier above the miniband energy level and by the width of the potential barrier. Therefore, the mobility in the superlattice direction of the lowest energy miniband can be made much smaller than the mobility of the conduction band in the same direction by increasing the width of the potential barrier between the wells.

On the other hand, the energy level of the lowest energy miniband in the potential well is determined by the width of the well, rather than the width of the barrier. Therefore, the dark current in the superlattice direction can be controlled by controlling the width of the CdTe layer, while the long wavelength cutoff of this type of superlattice when used as an infrared detector is determined by the width of the HgTe layer. As a result, the widths of the HgTe and CdTe layers can be varied to control independently the cutoff wavelength and the dark current.

This is not the case in a conventional $Hg_xCd_{1-x}Te$ detector, where the choice of x determines both the cutoff wavelength and the dark current. Thus the HgTe/CdTe superlattice, considered as a material, has one more degree of freedom than does $Hg_xCd_{1-x}Te$, and use of this degree of freedom would seem to allow the development of infrared detectors with performance improvements beyond what is available from the conventional HgCdTe technology.

3.3 TAILORABILITY AND TUNABILITY

The key idea of the doping superlattice, which is illustrated in Fig. 6, is that the number of donor atoms per period is approximately the same as the number of acceptor atoms per period. By the same mechanisms that were discussed in section 3.1, the fifth electron associated with the donor atoms will end up on the acceptor atoms in a neighboring region, and there will be a net positive charge in the donor layer and a net negative charge in the acceptor layer, but the structure will be electrically neutral overall. This gives rise to the modulation of the valence and conduction band edges as shown in the figure. The superlattice bandgap, that is, the offset between the lowest conduction miniband and the highest valence miniband, can be adjusted continuously up to the actual bandgap of GaAs by adjusting the magnitude of the donor and acceptor concentrations.

Notice that this bandgap is something new. It is an indirect gap in real space. Nevertheless, excitation and recombination can occur across this gap by tunnelling and other mechanisms. Since this gap can be varied from no gap to the gap of GaAs by varying the magnitude of the donor and acceptor concentrations in the superlattice, the doping superlattice represents the ultimate in tailorability.

Consider what happens if an electron is moved from a valence miniband to a neighboring conduction miniband. Some of the charge imbalance between the donor and acceptor layers is neutralized. This decreases the superlattice bandgap, and represents the ultimate in tunability.

This effect can actually be harnessed. The larger the period of the superlattice, the longer the lifetime of the miniband carriers before they recombine. This lifetime can be so long that the excited state is metastable over seconds. Devices that use incident radiation to create miniband electron-hole pairs and bias voltages to promote recombination have been constructed to allow the superlattice bandgap to be varied during the course of an experiment (Doehler 1986).

Photoconductors with a photoconductive gain in excess of 10^{11} have been also reported. These are very nonlinear, and it is not clear that they or similar devices will ever have any application in high-accuracy radiometry. In fact, the superlattice devices, such as detectors, amplifiers and nonlinear optical devices, that have been built so far have not been designed for radiometric applications, but for applications requiring a high gain-bandwidth product, such as optical communications. In these applications, the quantum-well superlattice devices have achieved levels of performance that are not possible with conventional materials.

Since the trade-offs encountered when maximizing radiometric accuracy are very different than those encountered when maximizing gain-bandwidth product, it is not surprising that efforts to develop devices optimized for the latter have not yielded noteworthy performance in the former area. However, the preceding discussion should make it clear that really amazing things can be done with quantum-well superlattices.

Of course, as in the past, it may be necessary for radiometrists to figure out how to exploit this new technology to solve problems in radiometry, because this field is probably too small and its applications too narrow to attract workers from the quantum-well superlattice field.

4. REFERENCES

Capasso F 1987 *J. Appl. Phys* **235** 172-176 and references therein.

Dingle R, Stormer H L, Gossard A C and Wiegmann W 1978 *Appl. Phys. Lett.* **33** 665-667

Doehler G H November 1983 *Scientific American* 144-151

Doehler G H 1986 *IEEE J. Quant. Elect.* **22** 1682-1695 and references therein

Petroff M D, Stapelbroek M G and Kleinhans W A 1987 *Appl. Phys. Lett.* **51** 406-408

Szmulowicz F and Madarsz F L 1987 *J. Appl. Phys.* **62** 2533-2540 and references therein to less accessible papers of Petroff M D and Stapelbroek M G

Inst. Phys. Conf. Ser. No. 92
Paper presented at Int. Conf. Optical Radiometry, NPL, London, 12–13 April 1988

Generalised radiance and practical radiometry

W T Welford
Blackett Laboratory, Imperial College
London SW7 2BZ, UK

Generalised radiance was proposed in 1968 by Adriaan Walther as a theoretical construct in wave optics to correspond to classical radiance in geometrical optics. It will be shown that the two do <u>not</u> correspond and that if radiance is measured experimentally the result is not predicted by the theory of generalized radiance.

On the definition of radiance for partially coherent sources

R. Martínez-Herrero and P. M. Mejías

Departamento de Optica, Facultad de Ciencias Físicas,
Universidad Complutense, 28040-Madrid, Spain

ABSTRACT: The general form of any generalized radiance
function of a partially coherent planar source is obtained
from the second-order coherence behavior of the source.
All these possible definitions of radiance are shown to be
consistent with the three basic requirements of the traditional radiometry. From the general expression, two recently
proposed definitions of radiance have been derived, one of
which is uniquely determined without arbitrariness from the
cross-spectral density function of the source. Also, for
completely incoherent sources, the generalized radiance reduces to the classical expression.

1. INTRODUCTION

As is well known, reexamination of basic radiometric concepts
such as radiance is required in order to extend the traditional radiometry to fields generated by planar sources of any
state of coherence. In fact, classical radiometry describes
the behavior of optical fields created by spatially completely
incoherent sources. However, it is also known that in any
radiating source the field is always correlated, at least over
distances of the order of the mean wavelength of the emitted
light. Note that, according to the Van Cittert-Zernike theorem
(Born and Wolf 1980), even a fictitious perfect thermal source
(e. g., a black-body source) can generate fields that are spatially highly coherent over arbitrarily large regions of space.
Therefore, the basic concepts of radiance, radiant intensity,
and radiant emittance should be generalized to fields generated by general stationary partially coherent planar sources
(primary or secondary).

Following the pioneering works of Walther (1968,1973), a generalization of such radiometric quantities was presented by
Marchand and Wolf (1972, 1974). However, they also showed that
in certain cases their radiance and radiant emittance may take
on negative values, in contradiction to one of the major postulates of traditional radiometry, and also exhibit other unphy-
sical features. Fortunately, the generalized radiant intensity
was always found to be nonnegative and to represent the angular distribution of the energy flux in the far zone, so this

concept retain the same physical meaning as that of the radiant intensity of classical radiometry.

To overcome these problems, two new definitions of radiance were recently proposed (Martínez-Herrero and Mejías 1984, 1986) that are non-linearly related to the cross-spectral density of the source and satisfy the three basic requirements of

i) being nonnegative,

ii) vanishing outside the source area,

iii) being the radiant intensity (calculated from the radiance) identical with the value obtained on the basis of the physical optics.

The former definition (Martínez-Herrero and Mejías 1984) is based in the following Mercer's expansion of the cross-spectral density $W(\bar{r}_1,\bar{r}_2)$ of the source (Gamo 1964, Martínez-Herrero 1979, Martínez-Herrero and Mejías 1981):

$$W(\bar{r}_1,\bar{r}_2) = \sum_n a_n^2 \, G_n^*(\bar{r}_1) \, G(\bar{r}_2) \tag{1}$$

where $\bar{r}_1$ and $\bar{r}_2$ denote points in the source plane z=0, and the coefficients a_n^2 and functions $G_n(\bar{r})$ are, respectively, the (positive) eigenvalues and eigenfunctions of the Fredholm integral equation

$$\int_D W(\bar{r}_1,\bar{r}_2) \, G_n(\bar{r}_1) \, d\bar{r}_1 = a_n^2 \, G_n(\bar{r}_2) \tag{2}$$

D being the finite region occupied by the source. It has been assumed that the source radiates into the half-space $z > 0$, and also, in order to simplify the notation, we have eliminated in all the expressions the explicit dependence on the frequency of the field. The radiance function $B(\bar{R},\bar{r})$ introduced by Martínez-Herrero and Mejías (1984) was defined in the form

$$B(\bar{R},\bar{s}) = \lambda^{-2}\cos\theta \left| \int_D \exp(ik\bar{s}\cdot\bar{r}) \, G(\bar{r},\bar{R}) \, d\bar{r} \right|^2 \tag{3}$$

where

$$G(\bar{r},\bar{R}) = C_D(\bar{R}) \sum_n b_n G_n(\bar{r}) \, G_n^*(\bar{R}) \tag{4}$$

with

$$b_n = \sqrt{a_n^2} \qquad\qquad (5)$$

$C_D(\overline{R})$ being the characteristic function of D. In eq.(3), λ is the wavelength of the light emitted by the source, $k=2\pi/\lambda$, $\overline{s}$ is the unit vector whose z component is nonnegative, and θ the angle between $\overline{s}$ and the z axis (see fig. 1). The radiance (3) was shown to be related to the energy flux according to the usual law of traditional radiometry and also to satisfy conditions i)-iii). However, it should be stressed that the generalized radiance given by eq.(3) really corresponds to a particular choice of coefficients b_n. In fact other generalized radiance functions satisfying requirements i)-iii) can be defined by choosing, for example,

$$b_n = a_n \exp(ih_n) \qquad\qquad (6)$$

where each h_n is real (Foley and Nieto-Vesperinas 1985). Equation (6) therefore gives rise to an infinite number of radiances of the form (3). In this connection, note that the coefficients b_n are not directly and uniquely determined from the second-order coherence characteristics of the source (a_n^2 and $G_n(\overline{r})$).

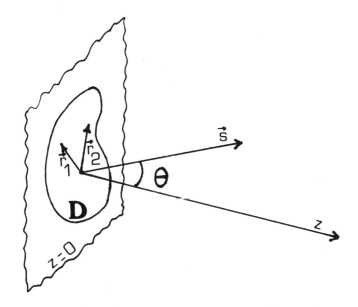

Fig. 1. Illustration of the notation used in this work

The purpose of the present paper is to go deeply into the possible definitions of radiance for partially coherent planar sources, analysing the general form that they should obey. It will be understood that a function $B(\bar{R},\bar{s})$ would represent a well-defined generalized radiance if it intrinsically fulfils properties i)-iii) for any source.

2.   GENERAL EXPRESSION OF GENERALIZED RADIANCE

As a first step in our work let us next show the following proposition (for the sake of brevity we omit demonstrations):

Proposition 1

Let $B(\bar{R},\bar{s})$ represent the generalized radiance function (according to an arbitrary radiance definition) of a partially coherent planar source. Then there exists a stochastic process $A(\bar{R},\bar{s})$ such that

a) $\langle A^*(\bar{R}_1,\bar{s})\, A(\bar{R}_2,\bar{s}) \rangle = \delta(\bar{R}_1 - \bar{R}_2)\, M(\bar{s},\bar{R}_1,\bar{R}_2)$

$$(7)$$

with $M(\bar{s},\bar{R},\bar{R}) \geqslant 0$

where $\langle \cdot \rangle$ denotes the ensemble average,

b) $\int_D \langle |A(\bar{R},\bar{s})|^2 \rangle\, d\bar{R} = \lambda^2\, (\cos\theta)^{-2}\, J(\bar{s})$ (8)

where $J(\bar{s})$ is the radiant intensity of the source, and

c) $B(\bar{R},\bar{s}) = \lambda^{-2}\, \cos\theta\, \langle |A(\bar{R},\bar{s})|^2 \rangle$.

$$(9)$$

In other words, a process $A(\bar{R},\bar{s})$ can be found satisfying a)-c) (from which the radiance is obtained) for any radiance definition. This implies that all possible generalized radiances are of the form (9).

For the sake of completeness the following proposition should also be included:

Proposition 2

Let $A(\bar{R},\bar{s})$ be any stochastic process such that

$$\int_D < | A(\bar{R},\bar{s}) |^2 > d\bar{R} = \lambda^2 (\cos\theta)^{-2} J(\bar{s})$$

where $J(\bar{s})$ is the radiant intensity of a general partially coherent planar source. Then the function $F(\bar{R},\bar{s})$ defined as follows

$$F(\bar{R},\bar{s}) = \lambda^{-2} \cos\theta \; C_D(R) < | A(\bar{R},\bar{s}) |^2 > \tag{10}$$

constitutes a well-established definition of generalized radiance.

Note that propositions 1 and 2 form, in a sense, a necessary and suficient criterion to define any radiance satisfying the three major postulates i)-iii).

Let us now recall the expression that gives the radiant intensity on the basis of physical optics (in terms of the cross-spectral density of the source):

$$J(\bar{s}) = \lambda^{-2} \cos^2\theta \iint_D \exp(ik\bar{s}\cdot(\bar{r}_2-\bar{r}_1)) \; W(\bar{r}_1,\bar{r}_2) \; d\bar{r}_1 \, d\bar{r}_2 \; . \tag{11}$$

Taking this equation into account, it can be seen at once that any stochastic process $A(\bar{R},\bar{s})$ fulfilling proposition 2 should satisfy

$$\iint_D \exp(ik\bar{s}\cdot(\bar{r}_2-\bar{r}_1)) \; (W(\bar{r}_1,\bar{r}_2) - \int_D <A^*(\bar{R},\bar{r}_1) \; A(\bar{R},\bar{r}_2)> d\bar{R})d\bar{r}_1 d\bar{r}_2$$

$$\tag{12}$$

$$= 0$$

where $\tilde{A}$ denotes the Fourier transform of A with respect to the variable $\bar{s}$. Now, out of the processes satisfying eq.(12) we will consider as physically acceptable those ones that are exclusively connected with the second-order behavior of the source itself, i.e., those fulfilling

$$\int_D \langle A^*(\bar{R},\bar{r}_1) \, A(\bar{R},\bar{r}_2) \rangle \, d\bar{R} = W(\bar{r}_1,\bar{r}_2) \quad . \tag{13}$$

The above statement neglects the functions belonging to the kernel of the integral operator $\hat{K}$,

$$\hat{K} \, f \equiv \iint_D e^{ik\bar{s}\cdot(\bar{r}_2-\bar{r}_1)} \, f(\bar{r}_1,\bar{r}_2) \, d\bar{r}_1 \, d\bar{r}_2 \tag{14}$$

since they contain no information at all about the physical characteristics of the particular source.

On the basis of the above assumption, any definition of radiance for partially coherent planar sources can be shown to obey the general expression

$$B(\bar{R},\bar{s}) = \lambda^{-2} \cos\theta \ C_D(\bar{R}) \sum_{n,m} \tilde{G}_n^*(\bar{s}) \ \tilde{G}_m(\bar{s}) \ H_{nm}(\bar{R}) \tag{15}$$

where $\tilde{G}_n$ are the Fourier transform of functions $G_n(\bar{r})$ (see eqs.(1) and (2)), and H_{nm} may be any function of the form

$$H_{nm}(\bar{R}) = \sum_i L_{ni}^*(\bar{R}) \ L_{mi}(\bar{R}) \tag{16}$$

satisfying

$$\int_D H_{nm}(\bar{R}) \, d\bar{R} = \delta_{nm} \ a_n^2 \tag{17}$$

where coefficients a_n^2 are defined in eqs.(1) and (2), L_{ni} are arbitrary functions and δ_{nm} is the Kronecker's symbol.

3.   SOME PARTICULAR CASES

From the general expression (15), other previously proposed definitions of radiance can be derived at once. Thus, if

$$H_{nm}(\bar{R}) = a_n \ a_m \ e^{i(h_m-h_n)} \ G_n^*(\bar{R}) \ G_m(\bar{R}) \tag{18}$$

we obtain the definition given by eq.(3) (see Martínez-Herrero and Mejías 1984).

On the other hand, if

$$H_{nm}(\bar{R}) = \delta_{nm} a_n^2 |G_n(\bar{R})|^2 \qquad (19)$$

we obtain the following definition of radiance introduced by Martínez-Herrero and Mejías (1986):

$$B(\bar{R},\bar{s}) = \lambda^{-2} \cos\theta \ C_D(\bar{R}) \sum_n a_n^2 |G_n(\bar{R})|^2 |\tilde{G}_n(\bar{s})|^2 . \qquad (20)$$

Notice that all factors that appear in eq.(20) are deterministic and can directly and uniquely be determined without arbitrariness from the second-order coherence features of the source.

Finally, if the source is assumed to be completely spatially incoherent then

$$A(\bar{R},\bar{s}) = i(\bar{R}) \ e^{ik\bar{s}\cdot\bar{R}} \qquad (21)$$

where $i(\bar{R})$ is a stochastic process such that

$$\langle i*(\bar{R}_1) \ i(\bar{R}_2) \rangle = \delta(\bar{R}_1-\bar{R}_2) \ I(\frac{\bar{R}_1+\bar{R}_2}{2}) \qquad (22)$$

$I(\bar{R})$ being the optical intensity at the source plane. The radiance in this case becomes

$$B(\bar{R},\bar{s}) = \lambda^{-2} \cos\theta \ I(\bar{R}) \qquad (23)$$

which is the well-known expression for the radiance of incoherent sources (Marchand and Wolf 1974).

It is important to note that Walther's definitions of radiance (Walther 1968, 1973) does not follow from the general expression (15) since they cannot be strictly associated with true radiance functions as we mentioned in the introduction.

REFERENCES

Born M and Wolf E 1980 Principles of Optics (Oxford: Pergamon Press)
Foley J T and Nieto-Vesperinas M 1985 J. Opt. Soc. Am. A **2** 1446
Gamo H 1964 Progress in Optics, Vol 3 ed E Wolf (Amsterdam: North-Holland) pp 187-336
Marchand E W and Wolf E 1972 J. Opt. Soc. Am. **62** 1972
Marchand E W and Wolf E 1974 J. Opt. Soc. Am. **64** 1219
Martínez-Herrero R 1979 Nuovo Cimento B **54** 205
Martínez-Herrero R and Mejías P M 1981 Opt. Commun. **37** 234
Martínez-Herrero R and Mejías P M 1984 J. Opt. Soc. Am. A **1** 556
Martínez-Herrero R and Mejías P M 1986 J. Opt. Soc. Am. A **3** 1055
Walther A 1968 J. Opt. Soc. Am. **58** 1256
Walther A 1973 J. Opt. Soc. Am. **63** 1622

Inst. Phys. Conf. Ser. No. 92

Paper presented at Int. Conf. Optical Radiometry, NPL, London, 12–13 April 1988

The manufacture and measurement of precision apertures

Miss T M Goodman, J E Martin, B D Shipp and N P Turner

NPL, Teddington, UK

ABSTRACT: The accuracy with which the area of an aperture can be measured is determined not only by the uncertainty in the measurement of any particular diameter but also by the quality of the aperture (eg. the departures from circularity). This paper discusses both these aspects, with specific examples of measurements made at NPL.

1. INTRODUCTION

The advances in the development of optical radiation detectors, both primary and secondary, have made it possible to measure radiant power with an uncertainty of about 0.01%. However, when a detector is used with an aperture system to measure radiant intensity, irradiance etc., this measurement potential cannot be fully utilised as the limiting uncertainty arises not from the detector itself but from the defining geometry. A circular aperture is normally used to define a detector or source area and a second aperture, aligned co-axially, is frequently used to define a solid angle of view; it is the uncertainty in the measurement of the aperture diameter, and how it relates to the mean diameter, that is becoming critically important. For example, to achieve an uncertainty of 0.01% in the area of a 10 mm diameter aperture (with perfect circularity), the diameter has to be measured with an uncertainty of 0.5 µm. Before it is possible to make these high accuracy measurements careful consideration must be given to the aperture design and manufacture.

This paper describes two techniques for making precision apertures and compares the non-contact method of measuring the aperture diameter using an optical microscope with a contact method using the NPL-designed internal-diameter measuring machine. The advantages and disadvantages of each type of aperture and measurement technique are also outlined.

2. CONSTRUCTION

2.1 Mechanical

Apertures have traditionally been made using normal workshop practices. The general procedure is for the aperture to be machine-turned or ground (if a hard material is chosen) to the approximate shape required and then hand-lapped to the desired size and surface finish. One future development, currently under investigation at NPL, is to diamond-turn the aperture to the final size.

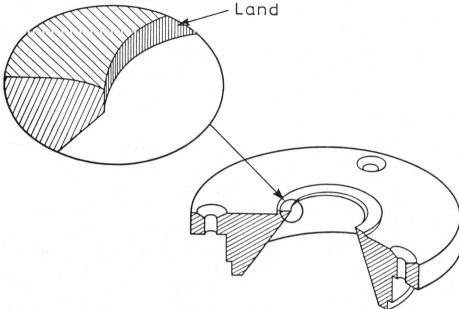

Figure 1. Schematic diagram of a mechanically-worked aperture, showing the position of the land.

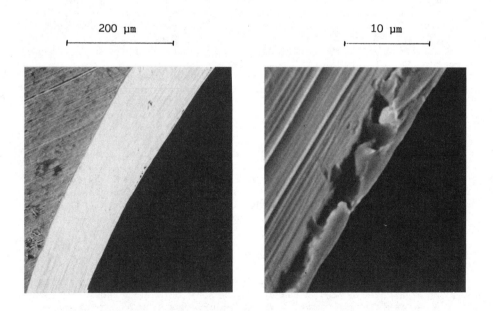

Figure 2. Scanning electron-microscope photographs of the land.

Advantages:

a) A variety of metals can be used to make the apertures, from soft copper to hardened steel. Hard materials are preferred as it is easier to produce sharp edges.

b) The edges of the aperture can be well defined, with the provision of a "land" as shown in figure 1.

c) The aperture diameter can be measured using either a contact or a non-contact method.

d) Because the apertures can be measured using a contact method (see section 3.1), it is possible to determine the circularity, thus enabling the area to be calculated with greater accuracy.

Disadvantages:

a) Quality apertures are very difficult to manufacture and are dependent on the skill of the craftsman.

b) The manufacturing time is considerable and hence the cost is high.

c) A lapping jig unique to each aperture is required and this also adds to the cost of manufacture.

d) The lapping process often throws up an unwanted burr at the edge, as shown in figure 2. Apertures must therefore be examined for burrs using a microscope (preferably a scanning electron-microscope).

2.2 Electro-deposition

Good quality circular discs of large dimensions are generally easier to produce mechanically than small area discs; for example, a departure from circularity of 5 μm on a 100 mm diameter disc corresponds to 0.5 μm on a 10 mm disc. This fact is exploited in the manufacture of apertures by electro-deposition. A "master" disc is produced, which is the same shape as the desired aperture but larger in size, and this is then used to generate a photo-resist mask of the required size by a photographic reduction technique. The mask is laid on a phosphor bronze substrate and copper or nickel is electro-deposited until the material just creeps over the sides of the mask. The photo-resist can then be dissolved and the aperture removed from the substrate (figure 3).

Advantages:

a) They are quick and cheap to manufacture, the production of the master disc being the most expensive and time-consuming process. Many apertures of different sizes can be produced from this one master.

b) They are commercially available. It is therefore not necessary to have access to workshop facilities and skilled craftsmen.

c) Apertures produced by this technique have a very shallow land.

d) The edges are smooth and without burrs, and sudden localised departures from the desired shape are unlikely.

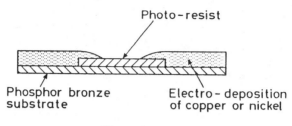

Photo-resist

Phosphor bronze substrate

Electro-deposition of copper or nickel

Stage 1

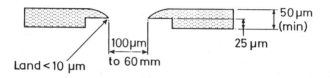

50 μm (min)

25 μm

Land < 10 μm

100 μm

to 60 mm

Stage 2

Figure 3a. Manufacture of apertures by electro-deposition.

Figure 3b. Apertures can be produced in a range of sizes.

e) It is possible to produce apertures in many different sizes and shapes (eg. circular, square, rectangular, from ~ 100 μm to 60 mm).

Disadvantages:

a) The circularity depends on the uniformity of deposition, which is difficult to control and guarantee.

b) The edges of the apertures are easily deformed under pressure, so these apertures are not suitable for measurement using a contact method, where a stylus loading is a necessary requirement.

c) The circularity is difficult to check.

d) The apertures are thin, and therefore fragile and easily distorted.

3. MEASUREMENT

3.1 Contact method

A special instrument has been developed at NPL for measuring internal diameters, using a contact method. The aperture circularity is first determined using a Talycenta roundness measuring machine, and diameters are then measured in two orthogonal planes (which can be related to the Talycenta trace - see figures 4 and 5). The diameters are measured using the NPL internal-diameter measuring machine, which is a precision comparator employing a laser interferometer measurement system based on a rigidly mounted stylus and a linear air bearing. The contact force of the stylus (0.06 N) on the aperture is highly reproducible. The calibration of the machine needed to give absolute values of diameter is established by using a silica box-standard which has already been separately and absolutely calibrated by white light interferometry, thus providing a direct link to the national standard of length. By combining the measurements of the diameter and circularity, a very accurate value for the area can be determined.

Advantages:

a) Any individual diameter can be measured with an uncertainty of better than 0.4 μm. (The machine is capable of measuring a grade AA plain setting ring of comparable size with an uncertainty of 0.2 μm.)

b) The circularity of the aperture can be accurately assessed.

c) Only one diameter measurement, with the Talycenta trace, is required in order to calculate the area of an aperture. As a cross-check, however, two diameters are normally measured.

Disadvantages:

a) This method cannot be used with knife-edge apertures. A minimum land of 70 μm is required against which the stylus can be located.

b) The presence of the land can cause problems in the use of the aperture. For example, corrections may have to be determined for radiation absorbed or reflected at this surface.

c) The minimum diameter that can be measured is 3 mm.

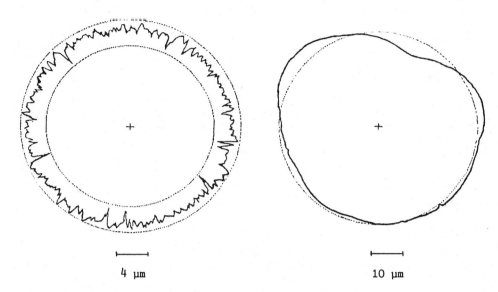

Figure 4. Talycenta traces of two poor apertures.

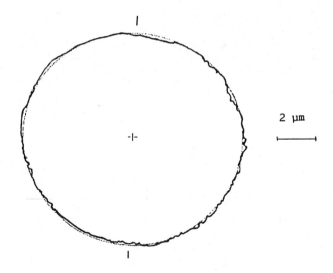

Mean measured diameters, expressed to the nearest 0.000 05 mm, are:

In plane of marked line .... 6.131 80 mm
At 90° to marked line .... 6.132 10 mm

Uncertainty at the 95% confidence level .... ±0.000 35 mm

Figure 5. Talycenta trace of a good aperture with the corresponding measured diameters.

3.2 Non-contact (optical) method

The diameter of an aperture can be measured using optical microscopy, by means of several different techniques. One method is to use a travelling microscope, whose motion can be measured with a laser interferometer. Alternatively, an image-shearing eyepiece can be used on a stationary microscope and calibrated against an interferometrically-measured stage micrometer. In the latter method, the maximum size of aperture which can be measured is determined by the finite shear distance within the field of view; larger apertures can be accommodated by reducing the magnification, but at the cost of correspondingly poorer resolution. It is, however, possible to use a special eyepiece on a travelling microscope, which shears the image to a degree determined by the position of the image with respect to the optical axis of the eyepiece. This form of shearing is not limited to the field of view of the microscope. In this case, rather than simply locating each edge of an image against a reference line in the eyepiece (filar setting, as used in a traditional travelling microscope) the adjustment is such that opposite edges of the two images just touch. Setting the edge by this method gives more repeatable results and is considered to have a lower systematic uncertainty than filar setting.

Advantages:

a) There is no possibility of damage to the edge. It is therefore suitable for the most fragile apertures.

b) Very large and very small apertures can be measured.

c) Any particular diameter can be measured very precisely. The uncertainty is typically less than 1.0 μm (see example in figure 6).

Disadvantages:

a) In order to determine the circularity of an aperture and accurately calculate its area, measurements must be made of a large number of diameters and care taken to ensure that these measurements are made across a common point - the nominal centre of the circle.

b) For the highest accuracy, a knife-edge is preferred. The presence of a land makes location of the edge more difficult, so this method is most suitable for apertures formed by electro-deposition techniques.

c) The accuracy with which the area can be determined may be limited by the edge quality. Any roughness at the edge, for example, makes it difficult to precisely determine any particular diameter, while local imperfections such as burrs may not be detected since only a limited number of diameters are measured.

4. CONCLUSIONS

Electro-deposited apertures, measured with an optical microscope, provide a quick, reliable and cheap method of defining an area with an uncertainty of about 0.05%. For the determination of an area to the highest accuracy, however, a contact method for determining the aperture diameter and circularity must be used in conjunction with a mechanically-worked aperture.

Repeated measurements on a single diameter using the non-contact method:

ZERO	LENGTH	ZERO
.01161	11.98717	.01139
.01110	11.98729	.01163
.01080	11.98729	.01141
.01108	11.98717	.01131
.01124	11.98720	.01124

1ST ZERO= .01117 2ND = .01140

AVERAGE ZERO = .01128

CORRECTED VALUE = 11.97594

ZERO	LENGTH	ZERO
11.98716	.01127	11.98720
11.98709	.01098	11.98709
11.98721	.01085	11.98712
11.98712	.01125	11.98718
11.98713	.01115	11.98720

1ST ZERO=11.98714 2ND =11.98716

AVERAGE ZERO =11.98715

CORRECTED VALUE =-11.97605

ZERO	LENGTH	ZERO
11.98768	.01161	11.98774
11.98776	.01139	11.98777
11.98774	.01121	11.98787
11.98783	.01126	11.98777
11.98774	.01107	11.98779

1ST ZERO=11.98775 2ND =11.98779

AVERAGE ZERO =11.98777

CORRECTED VALUE =-11.97646

ZERO	LENGTH	ZERO
.01071	11.98771	.01110
.01092	11.98770	.01117
.01099	11.98771	.01111
.01066	11.98764	.01106
.01087	11.98765	.01114

1ST ZERO= .01083 2ND = .01112

AVERAGE ZERO = .01097

CORRECTED VALUE = 11.97671

ZERO	LENGTH	ZERO
.01117	11.98793	.01143
.01148	11.98779	.01127
.01131	11.98803	.01147
.01147	11.98788	.01140
.01130	11.98797	.01137

1ST ZERO= .01135 2ND = .01139

AVERAGE ZERO = .01137

CORRECTED VALUE = 11.97655

ZERO	LENGTH	ZERO
11.98794	.01139	11.98794
11.98787	.01149	11.98804
11.98795	.01125	11.98800
11.98792	.01144	11.98804
11.98796	.01150	11.98789

1ST ZERO=11.98793 2ND =11.98798

AVERAGE ZERO =11.98796

CORRECTED VALUE =-11.97654

Mean value for this diameter .... 11.9764 mm
Uncertainty at the 95% confidence level .... ±0.0010 mm

Measurement across three other diameters gave:

11.9759 mm   11.9779 mm   11.9798 mm

Although the individual measurements are very precise, the overall uncertainty is determined to a large extent by the inability to measure the circularity accurately.

The overall uncertainty in the area of this 12 mm diameter aperture was estimated as ±0.05% at the 95% confidence level.

Figure 6. Non-contact measurements on an electro-deposited aperture.

Inst. Phys. Conf. Ser. No. 92
Paper presented at Int. Conf. Optical Radiometry, NPL, London, 12–13 April 1988

Total solar irradiance values determined using earth radiation budget experiment (ERBE) radiometers

Robert B Lee III

Atmospheric Sciences Division, NASA Langley Research Center
Hampton, Virginia 23665-5225 U.S.A

Michael A Gibson and Sudha Natarajan
ST Systems Corporation
Hampton, Virginia 23666 U.S.A

ABSTRACT: During the October 1984 through January 1988 period, the ERBE solar monitors on the NASA Earth Radiation Satellite and on the National Oceanic and Atmospheric Administration NOAA 9 and NOAA 10 spacecraft were used to obtain mean total solar irradiance values of 1365, 1365, and 1363 W/m^2, respectively. Secular variations in the solar irradiance have been observed, and they appear to be correlated with solar activity.

1. INTRODUCTION

A series of Earth Radiation Budget Experiment (ERBE) active-cavity type radiometers, called the solar monitors (Lee et al 1987), are the most recent pyrheliometers to be placed into space. The monitors were placed into orbit aboard the National Aeronautics and Space Administration (NASA) Earth Radiation Budget Satellite (ERBS) and the National Oceanic and Atmospheric Administration NOAA 9 and NOAA 10 spacecraft platforms during October 1984, December 1984, and September 1986, respectively. The monitors are used to measure the total solar irradiance. These measurements are then used as references in the calibrations (Luther et al 1986) of the three ERBE scanning and nonscanning radiometric packages which are found on each spacecraft platform. Once every 2 weeks, the Sun is observed by each monitor, almost simultaneously, for several 64-second measurement intervals. Each interval is separated into two 32-second periods in which the unocculted Sun drifts across the 13.7° field of view and its radiation field is measured, and in which a low emittance aluminum shutter, representative of a near-zero irradiance source, is cycled into the field of view and its radiation field is measured. Only the measurements taken during the last 4 seconds of a 32-second measurement period are used to derive the magnitude of the irradiance. The data reduction model normalizes the measurements to 1 astronomical unit, corrects for the radiation emitted by the active cavity and lost through the secondary aperture, accounts for the angular (cosine) responses of the monitors, accounts for variations in the areas of the primary apertures and electrical heater resistances with temperature, and corrects for differences in the thermal environment of the monitors between the shuttered and unshuttered measurement periods. The data reduction model is described in detail by Lee et al (1987). The measurement accuracy has been estimated to be 0.2%.

2.   DISCUSSIONS

For the October 25, 1984, through February 3, 1988, period, the solar irradiance values (normalized to 1 AU) derived from the ERBS, NOAA 9, and NOAA 10 solar monitor measurements are presented as functions of time in Figure 1. An ERBS or a NOAA 9 irradiance value represents typically the mean of two 64-second measurement intervals. For the NOAA 10, an irradiance value represents the mean of five intervals. The precession of the ERBS orbit allows the number of measurement intervals to rise as high as eight. The number of measurement intervals is influenced by the angular position of the Sun with respect to the orbital plane. For the precessing orbit of the ERBS spacecraft, the angular position changes approximately 5 degrees a day. In the case of the Sun-synchronous orbit of the NOAA 9 spacecraft, the Sun is approximately 38 degrees out of the orbital plane. For the Sun-synchronous orbit of the NOAA 10 spacecraft, the Sun is 67 degrees out of the orbital plane. The irradiance values presented in Figure 1 are available from the NASA Pilot Climate Data Center (Reph 1988). The mean irradiance values derived from the ERBS, NOAA 9, and NOAA 10 measurements were found to be 1364.9+0.2, 1364.7+0.5, and 1363.2+0.3 W/m^2, respectively. The corresponding measurement precisions were found to be 0.2, 0.5, and 0.8 W/m^2, and are indicated by the error bars in Figure 1. The absolute accuracy of the solar monitor measurements has been estimated to be 2.7 W/m^2 ($\pm$ 0.2%).

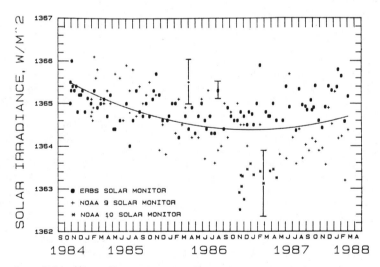

Figure 1.   Solar irradiance derived from the Earth Radiation Budget Satellite (ERBS), National Oceanic and Atmospheric Administration NOAA 9 and NOAA 10 spacecraft solar monitors. The solid line represents a second order polynomial fit through all three time series. The error bars represent the measurement precisions for each monitor. The absolute accuracy of each value has been estimated to be 2.7 W/m^2.

Although the NOAA 10 irradiance values are 1.7 W/m^2 lower than those of the ERBS and NOAA 9, the agreement among the three data sets is excellent. The NOAA 10 time series was limited to a 6-month period because its shutter mechanism failed on April 1, 1987. During ground tests, the NOAA 10 mechanism displayed some operational problems whereas the ERBS and NOAA 9 mechanisms exhibited no operational problems.

In Figure 1, the most obvious features in the irradiance time series are the secular decreasing trends before mid-1986 and the upward swing in the series after mid-1986, which are highlighted by a second order polynomial fit through the series. In examining the ERBS time series, the slope changed from a decreasing one of approximately -0.03% per year before mid-1986 to an increasing one of approximately +0.02% per year. In the case of the NOAA 9 time series, the slope changed from -0.05% to -0.02% per year. It is believed that these trends are correlated with solar activity. Using smoothed sunspot numbers as a guide, Koeckelenberg (1988) has estimated that solar cycle number 21 ended around September 1986, when solar cycle number 22 started. As the end of a cycle is approached, solar activity (sunspots, flares, faculae, etc.) decreases to a minimum and is accented by low numbers of sunspots on the photosphere. On the other hand, at the onset of a cycle, solar activity increases rapidly to maximum levels within 4 to 5 years. The sunspots are dark features on the photosphere. They have cooler temperatures than the surrounding photosphere and emit less energy. Consequently, when they drift across the surface of the Sun in large numbers, short-term decreases in the irradiance may be sensed by an orbiting pyrheliometer. Faculae are bright features on the photosphere, and they are found typically in the vicinity of sunspots. They emit more energy than the surrounding photosphere. The numbers of sunspots and faculae increase as maximum solar activity is approached and decrease as minimum activity is approached. Foukal and Lean (1988) suggest that the observed secular decreases in the Nimbus 7 (Hickey 1988) and Solar Maximum Mission (Willson et al 1986) irradiance time series between 1981 and 1984 may be related to the decreasing brightening effects of faculae as minimum solar activity is approached. Considering the Foukal and Lean hypothesis, it is possible that the observed ERBE secular solar irradiance trends may be linked to solar activity.

It is expected that the ERBS and NOAA 9 solar monitors will continue to yield valuable solar irradiance measurements throughout solar cycle number 22 and up to maximum activity for solar cycle number 23 into the year 2000.

3. REFERENCES

Foukal P and Lean J 1988 Proc. Solar Radiative Output Variation ed P Foukal (Cambridge, Massachusetts: Cambridge Research Instrumentation) pp 323

Hickey J R, Alton B M and Kyle H L 1988 Proc. Solar Radiative Output Variation ed P Foukal (Cambridge, Massachusetts: Cambridge Research Instrumentation) pp 189

Koeckelenberg A 1988 (January) Sunspot Bulletin (Bruxelles, Belgium: Sunspot Index Data Center) pp 4

Lee R B, Barkstrom B R and Cess R D 1987 Applied Optics 26 3090

Luther M R, Lee R B, Barkstrom B R, Cooper J E, Cess R D and Duncan C H 1986 Applied Optics 25 540

Reph M 1988 National Space Sciences Data Center, Code 634, NASA Goddard
   Space Flight Center, Greenbelt, Maryland 20771 U.S.A
Willson R C, Hudson H.S, Frohlich C and Brusa R W 1986 Science 234
   1114

Inst. Phys. Conf. Ser. No. 92
Paper presented at Int. Conf. Optical Radiometry, NPL, London, 12–13 April 1988

Comparison of CNES spherical and NASA hemispherical large aperture integrating sources, I, as determined with a laboratory spectroradiometer

B. Guenther, NASA Goddard Space Flight Center, Mail Code 673, Greenbelt, Maryland 20771 U.S.A. and M. Leroy and P. Henry, CNES Centre Spatial de Toulouse, Mail Code CT/APP/TI/AS 31055 Toulouse, France

The French national space agency, CNES, has purchased a large aperture integrating sphere to use in their calibrations of the SPOT instruments. This source was calibrated by the vendor at his manufacturing site and provided to CNES in the summer, 1987. The next SPOT sensors scheduled for flight are on platform number 2, and that platform was scheduled to be placed in storage in the late fall, 1987. Researchers from the CNES and the NASA Goddard Center participated in a calibration source comparison about the time of the SPOT-2 instruments were placed in storage. A Memorandum of Understanding was exchanged between the United States and French space agencies governing the activities of this cooperative agreement.

The US source used in this program is a 122 cm (four foot) diameter portable integrating sphere with a minimum aperture size of 25 cm and a maximum aperture size of 45 cm. The source had been transported to NASA facilities remote to the Goddard Greenbelt site.

The calibration of our source is accomplished in comparison to calibrated lamp standard of spectral irradiance through the use of a transfer spectroradiometer. This technique of source calibration has been used for US operational satellites for over a decade. The portable source was calibrated at 110 and 208 volts (for the lamp, power supplies and transfer spectroradiometer) before shipment to France, and the performance of the calibration equipment and source was verified when it was returned to the US. The stability of the hemisphere source was 1 percent for wavelengths between 0.55 and 1.1 microns.

At the calibrations in Toulouse, we had a problem with the operation of the lamp standard of spectral irradiance. The spectroradiometer response to the hemisphere source in Toulouse was within about two percent compared to what was measured in the US, but the response to the lamp standard was different by more than ten percent. Consequently we used the transfer spectroradiometer as a calibration standard. We obtained a Goddard calibration of the CNES source using the spectroradiometer, and achieved agreement of about two percent near 0.8 micron compared to the calibration provided by the sphere vendor.   Agreement between 0.4 and 0.7 microns was within five to eight percent. The uncertainty beyond 0.9 microns was larger due to a relatively large uncertainity in the wavelength scale of the transfer spectroradiometer. The transfer spectroradiometer used a silicon photodiode for these tests, and the actual wavelength registration of the monochrometer is important over wavelength ranges where the detector quantum efficiency changes rapidly with wavelength when the system is used as a calibrated artifact.   This experience demonstrates the importance of planning comparisons with duplicate and complementary techniques.

Inst. Phys. Conf. Ser. No. 92
Paper presented at Int. Conf. Optical Radiometry, NPL, London, 12–13 April 1988

Comparison of CNES spherical and NASA hemispherical large aperture integration sources, II, as determined with the SPOT 2 satellite instruments

M Leroy and P Henry

Centre National D'Etudes Spatiales

18 Avenue Edouard Belin

31055 Toulouse Cedex, France

B Guenther

Space and Earth Sciences Directorate

National Aeronautic and Space Administration

Goddard Space Flight Center

Greenbelt, MD 20771, USA

The absolute calibration, linearity and uniformity characteristics of a
CNES spherical and NASA hemispherical large aperture integrating calibration sources were compared at MATRA, TOULOUSE, France, in September,
1987, using a commercial laboratory spectroradiometer and the SPOT 2
satellite instruments. A Memorandum of Understanding between the CNES
and the NASA was written to define this project. This paper describes
the results of the source comparisons as determined with the SPOT 2
satellite instruments.

The angular uniformity of the sources in the field of view of the SPOT 2
instrument ($4°$) is tested by a comparison of the normalized responses of
the CCD detectors in the focal plane when illuminated by the sphere and
by the hemisphere. This comparison shows virtually no difference between
the responses, from which we can infer that both sources have an excellent
uniformity over a $4°$ field of view.

The linearity of the two sources has been investigated by plotting as a function of radiance the apparent calibration of SPOT 2 when seen by a given source (sphere or hemisphere) with a varying number of lamps on within the source. It is also possible to infer to some extent the SPOT 2 non linearity by using the data of non linearity of the two sources as seen by the spectroradiometer (see paper I).

Another important output of the experiment is the obtention of SPOT 2 absolute calibration coefficients by using the two sources with their own calibration. The paper describes the results and a tentative error budget to explain the 4 to 8% range discrepancies.

In summary, the results obtained by the SPOT 2 instruments can be considered as a validation of the comparison performed by the laboratory spectroradiometer and give also valuable insight into the range of accuracy of the characterisation of the SPOT 2 instrument performances (calibration, linearity, uniformity), using one given source.

Inst. Phys. Conf. Ser. No. 92
Paper presented at Int. Conf. Optical Radiometry, NPL, London, 12–13 April 1988

An optical heterodyne densitometer

A.L. Migdall, Zheng Ying Cong[*], J.E. Hardis

National Bureau of Standards, Gaithersburg MD 20899

Abstract: We are developing an optical heterodyne densitometer to measure optical density over an unprecedented dynamic range, with high accuracy, and high sensitivity. Our scheme uses a Mach-Zender interferometer with an acousto-optic modulator and an optical filter inserted into one of the interferometer arms. This method allows direct comparisons between optical and RF attenuators, ultimately tying optical measurements to RF attenuation standards. Using this technique, we have measured transmittances as small as 10^{-14}. We intend to move these measurements into the infra-red and to apply this technique to the problem of measuring wavefront distortion and scattered light.

1. Introduction

We have setup a system to measure optical density over an unprecedented dynamic range with high accuracy and high sensitivity. Using the ultra-sensitive technique of optical heterodyne detection, we have been able to measure optical attenuations of 10^7, 10^{10}, and 10^{12} at 633 nm with standard deviations of 0.5%, 2.5% and 20% respectively. In addition, we have begun recording changes in the wavefront curvature of the transmitted beams by taking angle resolved measurements.

Optical heterodyne detection (Boyd 1983) uses a strong reference laser beam, or local oscillator, to amplify very weak signals above the noise inherent in the detector. In this way, the shot noise of the strong reference beam can be made to dominate the measurement rather than the detector noise. Our setup (Fig. 1) uses a laser output split equally into two beams, one of which is sent through the filter under test and then frequency shifted by an acousto-optic modulator. The two beams are then recombined and sent to a detector producing a beat signal at the difference frequency. The amplitude of the beat signal is proportional to the square root of the filter transmittance. It is this square root dependance that, in part, gives heterodyne detection its increased dynamic range over ordinary densitometry measurements that depend linearly on transmittance.

This method makes possible optical transmittance measurements that are otherwise extremely difficult, the most obvious being accurate single frequency measurements of high attenuation absorbing glass filters.

Also, since this method is not limited to a specific optical frequency, a tunable laser source can be used to measure the transmission profile of an interference/blocking filter far out into the wings of the filter. Since the heterodyne efficiency depends on how well the two beams are overlaped or mode-matched, distortion of the transmitted beam can be seen in an angular map of the heterodyne signal. In addition, very low level scattering by optical components can be quantified.

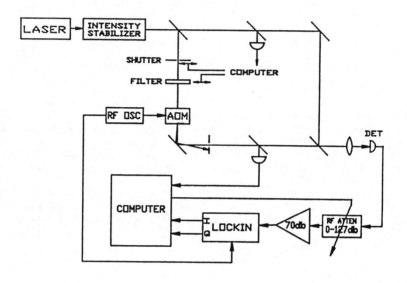

Figure 1   Schematic of Experiment

2. Theory

Optical heterodyne detection uses a "reference" light wave (beam 0) of one frequency to detect an optical signal (beam 1) at a second, slightly different frequency. The photocurrent, $i(t)$ from a photodetector is proportional to the light intensity incident on the device at time t and integrated over the detector area. Intensity is the square of the sum of the electric fields of the two beams E_0 and E_1, so the intensity at a point (x,y) is given by

$$I(x,y,t)=|E_0(x,y,t) + E_1(x,y,t)|^2.$$

If the two beams have different optical frequencies, ω and $\omega+\Delta$, then when $I(t)$ is averaged over many optical cycles, a cross term at the difference frequency Δ arises (in addition to a constant term). This term at the difference frequency is proportional to $E_0 E_1$ or $(I_0 I_1)^{1/2}$, where I_0 and I_1 are the intensities of the two individual beams. Thus, the square of the amplitude of the beat signal is proportional to the intensity of each beam. One can see that if the two beams differ in intensity, the more intense beam effectively amplifies the signal due to the weak beam.

When integrated over the detector area and averaged over an optical cycle, the signal current from the detector may be written as:

$$i(t) = H K [1 + 2(TP_1/P_0)^{1/2} \cos(\Delta t)]$$

where H is the heterodyne efficiency containing the integral of the spatial dependence of the two beams (described below), $K = P_0 e\eta/\hbar\omega$, P_0 and P_1 are the laser powers in the two beams, e is the electron charge, η is the detector quantum efficiency, $\hbar$ is Planck's constant divided by 2π, ω is the optical angular frequency, Δ is the difference frequency, and T is the transmittance (assumed to be <<1) of an optical filter inserted into the P_1 beam path. Thus, the optical attenuation is found by taking the square of the ratio of the beat signals obtained with and without the filter inserted.

It is important to note that this technique does not place extreme requirements on detector linearity, because the unattenuated beam effectively biases the photodetector. One can see that as the filter transmittance changes from full transmission to total attenuation the average intensity on the detector changes by only a factor of two.

The sensitivity of this technique may be defined as that transmittance that will produce a signal amplitude just equal to the noise in the system (ie. signal/noise = 1). If the unattenuated laser beam is sufficiently powerful, its shot noise will dominate the detector noise, so the noise current squared may be written as:

$$<i_n^2> = 2P_0 e^2 \eta B/\hbar\omega,$$

where B is the bandwidth of the measurement. The sensitivity is found by setting this equal to the heterodyne signal current squared written as:

$$<i_s^2> = 2e^2 \eta^2 H^2 TP_0 P_1/(\hbar\omega)^2$$

and solving for T. This gives a transmittance sensitivity $T=\hbar\omega B/\eta HP_1$. For a HeNe laser power of 1 mw in each beam, 100% detection efficiencies, and a 1 Hz measurement bandwidth this minimum detectable transmittance is 3×10^{-16}.

The heterodyne efficiency, H is the integral over the detector area, A, of the intensity at the beat frequency due to the two beams and is given by

$$H = \int_A E_0(x,y) E_1(x,y) dx dy .$$

This integral has been calculated for two Gaussian beams with given radii and whose centers coincide at the detector surface, one of which is incident normal to the surface, while the second beam is incident at angle α. The result of that calculation yields the functional form of H for small α as

$$H \propto \exp[-w^2 \alpha^2]$$

where w depends on the diameter of the two beams. This dependance should allow changes of the spatial structure of the transmitted beam to be observed. This will be the object of future study.

3. Experimental Technique

Our optical setup is based on a simple Mach-Zender interferometer where a 1 mw HeNe laser beam is split into two beams and subsequently recombined. The filter of interest is placed in one of the beams. The beam transmitted by the filter is then frequency shifted by an acousto-optic modulator (AOM) driven at 30 MHz. The unattenuated beam and the attenuated and frequency shifted beam are recombined by a beam splitter and focused onto a detector/amplifier. The detector produces a DC signal plus the AC component at the difference frequency of the two beams. The amplitude of the beat signal depends on the intensity of the recombined beam and thereby on the transmittance of the filter.

The AC output of the detector/amplifier is put into a computer controlled 0-127 db RF attenuator followed by two 35 db gain RF amplifiers and one or two crystal bandpass filters. This signal along with the AOM drive frequency is then fed into a 50 MHz lock-in amplifier. This produces two DC signals, one proportional to the beat signal in phase with the drive frequency and the other proportional to the signal in phase with the drive frequency shifted by 90 degrees. These inphase and quadrature signals are recorded by the computer which calculates the magnitude of the beat signal.

When the filter is removed from the optical path, the RF attenuator is increased to maintain a nearly constant signal at the lock-in. By alternately recording the beat signal (including the RF attenuation) with the filter in and out, any effects due to slow changes of the RF gain are eliminated. The computer also monitors the laser power and the power of the frequency shifted light exiting the AOM to normalize the beat signal. An optical shutter placed in the frequency shifted arm and controlled by the computer is used to zero the electronics.

This precise arrangement of components was chosen to minimize ordinarily small effects that become important at our high detection sensitivities. We found that the exit surface of the AOM backscattered light at the 10^{-9} level. This produced a beat signal even with the frequency shifted beam blocked, indicating that light backscattered by the output surface of the AOM reached the laser and was reflected into the unshifted beam. To prevent this, the test filter was placed upstream from the AOM so that the light that is both backscattered and frequency shifted is reduced twice by the filter attenuation. Several measures are taken to make the optical phase difference of the two beam paths stable to allow for very narrow bandwidth settings on the lock-in amplifier, and thus reduce the noise. The optical apparatus is enclosed in a box to reduce air turbulence, and the optical table is placed on vibration isolation legs.

Measurements are typically made with 1 s integration times on the outputs of the lock-in amplifier and the laser power monitors. A 10 second wait time is used to allow for these signals to settle after changing the shutter or filter position. These delays and the time to reposition the filter (currently via a very slow translator) result in a single measurement taking about 1.5 minutes.

4. Results

Tests have been made to characterize the noise and drifts associated with the technique. We found that individual measurements of moderate transmittances could be made with 1.5% standard deviation once the interferometer had stabilized. This was accomplished only after correcting for a systematic error of about 6% in the lock-in that varied with input phase angle. This correction reduced the problem to approximately .5%. We hope to reduce this further by designing and constructing our own lock-in.

We have made initial tests of the dynamic range and found that measurements of attenuations as great as 10^{-12} could be made with a standard deviation of 20% between individual measurements. We also measured filters with transmittances as low as 10^{-14} with the noise becoming roughly equal to the signal. This implies a sensitivity in the 10^{-14} range, leaving 2 orders of magnitude, before running up against the theoretical limit calculated from our current experimental parameters. Measurements at very high attenuations require very careful RF methods to assure that leakage paths do not add systematic errors to the RF attenuator calibration. We are currently working to improve our RF apparatus.

Preliminary measurements of the angular dependence of the heterodyne efficiency have been made. These results have been fit, with some success, to the functional form shown above. While the results of these fits are not yet quantitatively reliable they are in qualitative agreement with our calculations and will be studied further.

While most of our tests were made using a frequency stabilized HeNe laser, we have successfully run a test using a multimode Ar^+ pumped dye laser, demonstrating the tunability of the technique.

5. Conclusion

Our tests thus far have succeeded in demonstrating the extraordinary dynamic range and sensitivity of heterodyne measurement of optical density. This range far exceeds the capabilities of the ordinary attenuation measurements presently in use at NBS. Since the current measurements are not at the fundamental limits of the technique, further work should allow us to push the sensitivity at least an order of magnitude lower.

Planned improvements to this system include construction of a new lock-in amplifier to reduce systematic errors with respect to phase angle to below the current .5% level. We intend to produce accurate absolute measurements by completing the connection of our optical measurements to high precision RF electrical standards. 30 MHz was chosen for our beat frequency with this in mind, because this is where RF attenuators can be calibrated by an available service at NBS in Boulder. When this is done, we will compare our heterodyne measurements to conventional measurements where the dynamic ranges overlap.   In addition, tests will be done to verify the linearity of our technique.   Work is also underway to extend this technique to the infrared using a CO_2 laser at 10.6μm.

We intend to develop further the transmittance versus angle measurements to allow the total transmittance of a filter and the wavefront distortion to be quantified.   Another application where this high dynamic range technique should prove useful is in measuring the low angle scattering of both transmissive and reflective optics.   This should be particularly useful in the case of the low scatter super-polished mirrors now becoming available.

We greatfully acknowledge support for this work by the U.S. Army Strategic Defense Command.

R.W. Boyd, <u>Radiometry and the Detection of Optical Radiation</u>, p. 195, John Wiley & Sons, New York, 1983.

* Permanent address National Institute of Metrology, Beijing, People's Republic of China

Inst. Phys. Conf. Ser. No. 92
Paper presented at Int. Conf. Optical Radiometry, NPL, London, 12–13 April 1988

Ultraviolet radiometry

Henry Lyall

NEI International Research & Development Co.Ltd.

1.BACKGROUND

The majority of the work described here was carried out during the development of an ultraviolet radiometer for the European Space Agency (ESA).Some further data has arisen out of studies in conjuction with Newcastle Polytechnic involving students working on MSc. projects in Optoelectronics.Some of the original work has been described in detail elsewhere (1) and this presentation summarises the original work and further developments.

2.RADIOMETER SPECIFICATION

The ultraviolet radiometer required by ESA was intended for monitoring the ultraviolet content of essentially Air Mass 0 (AMO) radiation in a space simulation chamber over the temperature range from −20 to +80C. The specification called for a compact (36x72x4mm) detector head which would measure a continuous light source without chopping in seven bands in the UV with less than 10% out of band contribution. Although required primarily for measuring rather high irradiances (up to 200mWcm^{-2}) the specification also called for the ability to measure down to 1nWcm^{-2}. The measurement bands are shown in Figure 1 together with the AMO spectrum.

Examination of the spectrum indicated that considerable problems could be anticipated in optical filtering at the shorter wavelengths. Also calculations based on reasonable estimates of photodiode and filter characteristics showed that it would be necessary to accurately measure signal currents of the order of a picoamp to meet the specification.

3.CHOICE OF PHOTODETECTOR

The compact dimensions of the head and in particular the thickness of 4mm (later relaxed to 4.5mm!) dictated that only solid state photodetectors could be considered. At the time of the initial studies the choice lay between UV sensitive silicon photodiodes or gallium arsenide phosphide (GaAsP) photodiodes. Since the radiometers were built gallium phosphide (GaP) photodiodes have become available and are another alternative.

Representative types of Si photodiodes have been studied together with GaAsP and GaP. The samples selected for study are listed below and the responsivities at shorter wavelengths are shown in Figure 2.All of the Si photodiodes had a long wavelength response extending to about 1000nm which must be taken into consideration when choosing suitable optical filters.

Silicon diffused junction:	Centronic OSD 50-1
	EG&G UV215B
Silicon inversion layer:	UDT-UV50
GaAsP Shottky barrier:	Hamamatsu G1127
GaP Shottky barrier:	Hamamatsu G1962

4.PHOTODIODE STABILTIY

For this application it is important that the responsivity be stable both under high levels of irradiation and with respect to time.

The Centronic photodiode showed signs of instability at very low levels of UV and was not tested further. High level tests at 60mWcm^{-2}using the 254nm mercury line in a dry nitrogen atmosphere with no window over the photodiodes gave the results shown in Figure 3. All the silicon photodiodes showed considerable changes in responsivity, whilst the GaAsP and GaP photodiodes were much more stable. It is interesting to note that subsequent tests have shown that GaAsP and GaP photodiodes also degrade significantly if tested in packaged form with windows as normally supplied.

Silicon photodiodes also exhibit long term changes in UV responsivity without exposure to UV (2) as shown in Figure 4 whereas tests on about ten GaAsP photodiodes showed no discernible changes in responsivity over a period of a year. No long term tests have yet been carried out on GaP although one could perhaps be optimistic in view of the similarity to GaAsP both in construction and general performance.

With regard to silicon, EG&G (3) have indicated that they have indentified the problem of instability in the UV as being due to ionic contamination during manufacture, and that their present processing procedures avoid this problem.

5.ELECTRICAL MEASUREMENTS ON PHOTODIODES

In this application it is necessary to operate down to zero frequency and it is therefore important that there are no systematic offsets. Since the reverse saturation current, Io, which is strongly temperature dependent, may be much larger than the signal current, the photodiode must be operated at zero bias and this is therefore the point at which device parameters must be determined.

Manufacturers data on photodiodes is often inadequate, but fortunately sufficient data for design purposes can be obtained froma few simple tests. The photodiode is measured in the dark and is treated as an ordinary diode satisfying the diode equation:

$$I = Io*(exp(V/(nkT/q)-1)$$

where I is the current flowing as a result of a bias voltage V, Io is the reverse saturaton current and n is the ideality factor of the order of unity.

Io can be obtained from an extrapolation of a plot of the forward bias voltage V against logI, a curve which also gives some indication of the currents at which nonlinearity can be expected, when there is a deviation from the ideal form at high currents.This method gives results which agree well with direct measurements of the slope resistance at zero bias provided there is a well defined linear region. In the case of GaAsP and GaP it has the advantage of avoiding the time consuming measurement of currents of <100fA.

Figure 5 shows that GaAsP and GaP have very much lower values of Io than any of the silicon photodiodes, and that GaAsP conforms to the diode equation to much higher currents than the other photodiodes, consistent with its excellent linearity at high photocurrents.

Junction capacitances at zero bias were also measured and the results are summarised in the table below. The low values of reverse saturation current and high values of junction capacitance for GaAsP and GaP should be noted.

DEVICE	AREA mm^{-2}	Io	Io/mm^{-2}	%Io/K	C	C/mm^{-2}
OSD50	50	2nA	40pA		470pF	9.4pF
UDT UV50	50	370pA	7pA		2.8nF	56pF
EG&G UV215B	23.4	140pA	6pA	12	530pF	22pF
GaAsP	21.6	20fA	1fA	12	7nF	350pF
GaP	5.3	10fA	2fA	12	1.6nF	300pF

6.OPTICAL FILTERS

All of the photodiodes respond to a wide range of wavelengths and optical filters must be used to define the measurement bands shown in Figure 1. In view of the small fraction of the total irradiance which must be measured at the shorter wavelengths, the filters must have very good rejection outside the measurement band. The rejection at longer wavelengths is the most important, particularly if silicon photodiodes with a response extending into the infrared are to be used.

Three types of filter were considered for this application: all–dielectric interference filters; metal dielectric filters and glass absorption filters. Since all of the UV transmitting glasses show a high transmission above 700nm the latter were only relevant for use with GaAsP and GaP photodiodes.

The stability of the three types under high UV irradiance was studied with typical results shown in Figure 6.

The all–dielectric filters degraded badly, probably due to the use of materials such as zinc sulphide in their construction. The metal dielectric filters showed a small initial fall in transmission but were stable thereafter, and the glass absortion filters behaved in a similar manner.

Although not providing such a sharply delineated band as the best all–dielectric filters, a single two cavity metal dielectric filter provided adequate out of band rejection for most of the wavebands when used with GaAsP or GaP photodiodes. For the shortest waveband centred on 200nm it was necessary to cascade two such filters.

7. NOISE CONSIDERATIONS

The fundamental limit to the measurement of the currents from the photodiodes at zero bias is the thermal noise in the slope resistance. In all cases the increase in Io with temperature implies that this limit will deteriorate by a factor of two for every twelve degrees increase in temperature. In the case of a silicon photodiode such as the EG&G UV215B the slope resistance is about 100M at ambient temperature giving a limit in th region of 100fAHz $^{0.5}$ p–p and this limit can be approached fairly closely at low frequencies using commercially available amplifiers. In the case of GaAsP and GaP extrapolation indicates that the limit is about two orders of magnitude better, say in the region of 1fAHz $^{-0.5}$ p–p. There are however two significant obstacles to approaching this limit in a simple system. One is the requirement for very high values of feedback resistors ($>10^{12}$ohms) which have problems with long term stability and high temperature coefficients. The other, which is the dominant factor in practice, is the interaction of the noise voltage of the amplifier with the capacitative impedance of the photodiode. Using the most appropriate amplifier identified, the observed noise using 4.6mm square GaAsP photodiodes was typically 30fA p–p in an 0.25Hz bandwidth. Owing to the low initial value of Io this was only slightly degraded at elevated temperatures. The lowest noise observed with a system of this type is 10fA p–p in an 0.25Hz bandwidth using a 2.3mm square GaP photodiode.

8.INSTRUMENT DESCRIPTION

Two instruments were constructed for ESA based on the information obtained during a feasability study. GaAsP photodiodes were selected as being superior to silicon in this application with regard to linearity, UV stability, noise performance (particularly at elevated temperatures) and a more convenient long wavelength.If a similar instrument was to be built today GaP would be a better choice since it combines excellent UV stability with an even more convenient long wavelength cut–off with regard to filtering requirements

Glass absorption filters and two–cavity metal dielectric filters were used as appropriate and pre-aged before calibration. Only on the shortest waveband was it necessary to cascade two metal dielectric filters to achieve the required inband/out of band ratio.

The heart of the readout unit was a varactor bridge amplifier chosen for its combination of moderately low voltage noise and very low bias current (<10fA). Feedback resistors of up to 100G were used and a measurement bandwidth of either 2Hz or 0.25Hz could be selected according to requirements. A digital thermometer was incorporated to indicate the temperature of the head to allow the (small) temperature dependent corrections to be applied as necessary.

The detector heads were calibrated at normal incidence by the National Physical Laboratory for responsivity around the nominal measurement bands. This data was combined with out of band measurements and weighted against the AMO spectrum to derive the calibration in power density within the various wavebands.

ACKNOWLEDGEMENTS

The author wishes to thank ESA for permission to publish information on this project; Dr.A.D.Wilson (now at Barr & STroud,Glasgow) who as project leader in the first phase carried out many of the optical measurements assisted by Mr.R.F.Lee; and Mr.J.P.Lloyd who did much useful work on GaP as part of an MSc. project in Optoelectronics at Newcastle Polytechnic.

REFERENCES

1.A.D.Wilson and H.Lyall "Design of an ultraviolet radiometer.1.Detector electrical characteristics" &"Design of an ultraviolet radiometer.2.Detector optical characteristics" Appl. Opt 25,4530 & 4540 (1986).
2.P.J.Key and N.Fox: Personal communication.
3.J.Melnyck (EG&G): Personal communication.

Note: Reference 1 contains an extensive bibliography relating to this subject.

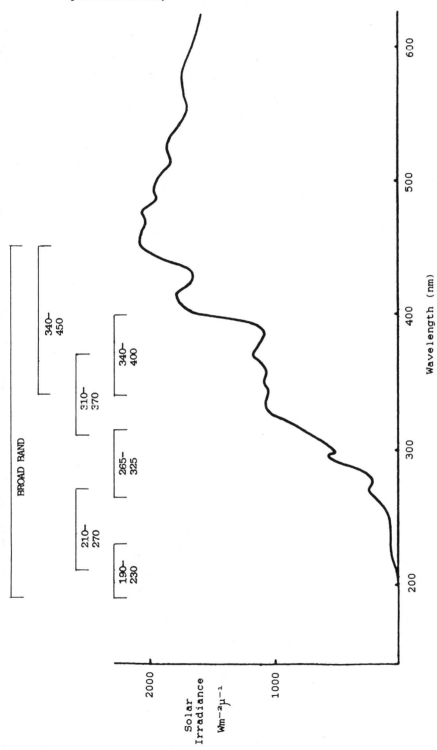

Figure 1. Air Mass 0 (AM0) spectrum and radiometer measurement bands.

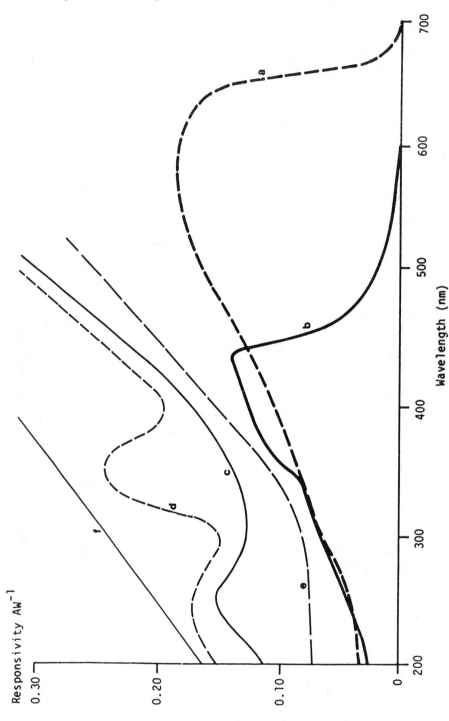

Figure 2.Responsivity versus wavelength
(a) GaAsP G1337. (b) GaP G1962. (c) EG&G UV215B. (d) UDT UV50.

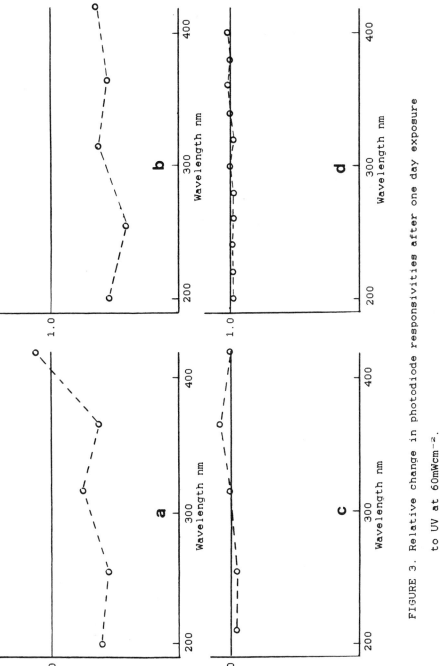

FIGURE 3. Relative change in photodiode responsivities after one day exposure to UV at 60mWcm^{-2}.

(a) UDT OSD 50 ; (b) EG&G UV215B ; (c) GaAsP G1227 ; (d) GaP G1962.

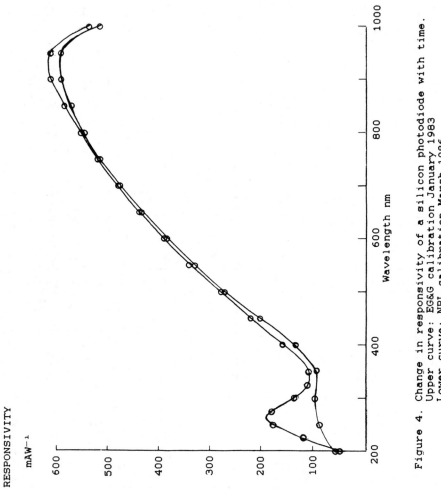

Figure 4.   Change in responsivity of a silicon photodiode with time.
Upper curve: EG&G calibration January 1983
Lower curve: NPL calibration March 1986

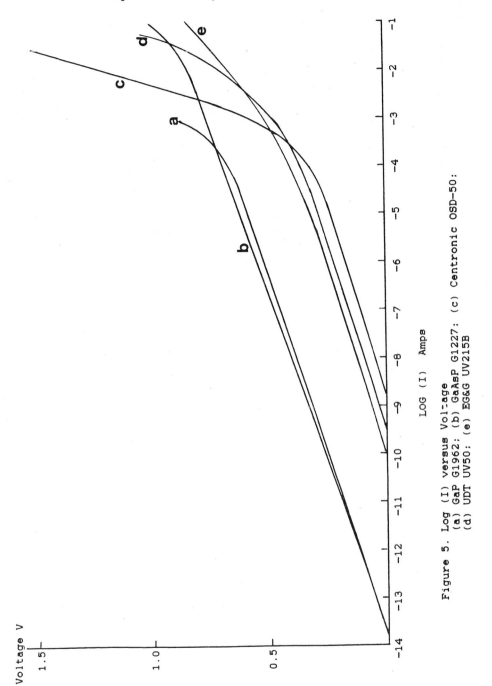

Figure 5. Log (I) versus Voltage
(a) GaP G1962; (b) GaAsP G1227; (c) Centronic OSD-50:
(d) UDT UV50: (e) EG&G UV215B

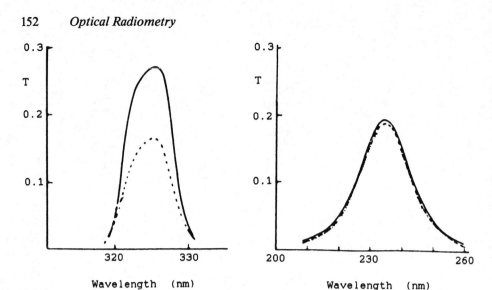

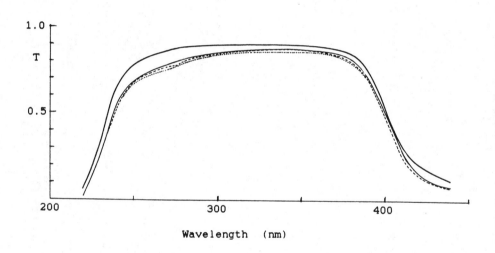

Figure 6.

Effect of irradiation at 60mWcm⁻² on the transmission T of filters.
(a) All-dielectric interference filter : --- 3hrs;
(b) Metal-dielectric interference filter: --- 6hrs;... 13 & 27 hrs.
(c) Glass absorption filter:—— 3hrs;--- 6hrs;... 13 & 27 hrs.

Inst. Phys. Conf. Ser. No. 92
Paper presented at Int. Conf. Optical Radiometry, NPL, London, 12–13 April 1988

Intercomparison of absolute radiometers at several visible laser wavelengths

J. Campos, A. Corróns, A. Pons and P. Corredera

Instituto de Optica "Daza de Valdés"
28006 Madrid - Spain

ABSTRACT: In this paper, a comparison between the ac
curacy of the measurements of the radiant flux of a⁻
laser beam at several wavelengths in the visible --
range with different absolute radiometers is shown.
The radiometers used are: an E.C.P.R. Rs 3940 from
Laser Precision Corp., a N.P.R.L. serie 2000 and an
UDT model Q.E.D. 100. Silicon photodiodes, which ab
solute spectral responsivity was determined by the⁻
self-calibration method, have also been used.

INTRODUCTION

Absolute radiometers are widely used at the present --
time not only to realize radiometric scales, but to measure
radiant power of many other applications where absolute values
are needed. There are different types of absolute radiometers
based on different materials and measurement principles. All
of them are supposed to be accurate in 1% or better, but in   -
measuring radiant power some other uncertainty sources, such
as sensitive surface in homogeneities, may increase the ultimate error of the measurement.

Since three different absolute radiometers are available
in the Laboratory of Radiometry at the Institute of Optics, it
would be interesting to measure the optical power of a light
beam with every radiometer under the same environmental conditions and to compare the radiant flux measured with every one.

This is the object of this work, which have been done
at seven laser wavelengths within the spectral range from 400
nm to 700 nm. Silicon photodiodes, which absolute spectral responsivity was determined by the self-calibration method, have
also been used in this work. The results obtained with them -
will be also included in the comparison.

DESCRIPTION OF THE RADIOMETERS

A brief description concerning to measurement principle
and structure will be given in this section for the radiometers
used in this work. Manufacturer and model details of the radiometers are given just for indentification purposes.

   A more complete and detailed description may be found in the papers referenced below, where the first development of -- these radiometers is described.

A) Electrically calibrated pyroelectric radiometer, model Rs 3940, serial number Rs 355, from Laser Precision Corp.. This radiometer uses the equivalence of radiant and electri cal heating. The radiant power is measured by generating an identical heating with controllable and measurable electri- cal power. The absorbing element is a flat black plate and the sensor is a pyroelectric detector.

   The first development and realization of this radiometer was shown by Blevin and Geist (1972,1973).

B) N.P.R.L. serie 2000 absolute radiometer. This radiometer uses a bolometer type detector in a bridge configuration. Its sensitive surface is coated with a black film and is located in a meridional plane of a hemispherical mirror, which returns back to the detector the reflected -- power.

   The measurement principle is also the equivalence of radiant and electrical heating. In this case the detector is preheat ed by an electrical current in absence of optical radiation. When radiation is present, the electrical current will change in order to balance the bridge again.

   Optical power is calculated from the difference on electri- cal power between the shuttered and unshuttered states. The first realization of this radiometer was shown by Hengst- berger (1977).

C) Q.E.D. 100, serial number 152, from United Detector Tech.: Absolute detector which responsivity is based entirely on known physical constants. The device is formed by four inver sion layer silicon photodiodes in a light trapping configu- ration.

   Incoming optical power is calculated from the photocurrent measurement. The original design of this radiometer was -- shown by Zalewski and Duda (1983).

   As previously mentioned, silicon photodiodes model UV- -444 B from EG & G, have also been used. Their absolute spec- tral responsivity was determined by the self-calibration meth- od (Geist et al. 1980), which is based on the evaluation of ex ternal quantum efficiency of these detectors. In this case, op tical power is calculated from the photocurrent measurement.

EXPERIMENTAL SET UP

   The light source used has been a Kr laser. The radio- meters used have different detectivities and saturation limits. Therefore the radiant flux coming into the detector had to be carefully selected. The radiant flux used at different wave- lengths was always in the range from 150 μW to 200 μW, which is within the linearity range of all the radiometers.

The laser source has been spatially filtered and colli-
mated, in order to obtain a more uniform spatial power distribution and to prevent the sensitivity of some radiometers to
the divergence degree.

The laser beam power has been monitored during the measurement process. This is specially important in this case, because the response time and the time needed to perform the mea
surement are very different for each one of the radiometers.

Finally, just to note that measurements where done under
normal incidence conditions, using a beam of four millimeters
in diameter and in laboratory temperature of 20.0°C + 1.0°C.
The laser beam was linearly polarized with the electrical vector normal to the plane of incidence.

RESULTS OBTAINED AND DISCUSSION

In Figure 1 are shown the results obtained for E.C.P.R.,
N.P.R.L. and Q.E.D. 100, and in Figure 2 those for three silicon photodiodes, identified as A-1, A-3 and PAC-2. These Figures show the relative dispersion of the radiant flux measurement versus wavelength for every radiometer. The relative dispersion is defined as the quotient between the flux measured
by every radiometer and the mean of the radiant fluxes measured
by the radiometers E.C.P.R., N.P.R.L. and Q.E.D. 100. The -
straight line on both Figures represents this average flux.

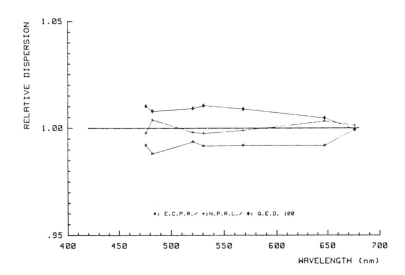

Figure 1.- Relative dispersion of the radiant flux measurement
for every radiometer.

RADIANT FLUX MEASUREMENT

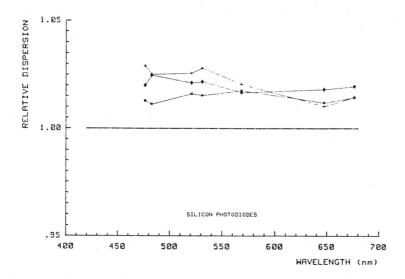

Figure 2.- Relative dispersion of the radiant flux measurement for the silicon photodiodes.

The results obtained with silicon photodiodes are not - used to calculate the average flux, because, as it is known - (Stock, 1987) silicon photodiodes present spectral aging pattern. The responsivity of silicon detectors used here was deter mined four years ago. This is a sufficiently long period of - time as to expect that their responsivity have changed.

The percent measurement precision is presented in Table I. Usually this is in the order of 0.1% or even lower for sili con detector based radiometers. This means that the differences observed in the Figures cannot be explained by the measurement precision. Therefore those differences seem to be systematic.

Table I: PERCENT MEASUREMENT PRECISION

Wavel.(nm)	E.C.P.R.	NPPL A.R.	Q.E.D.	A-1	A-3	PAC-2
476.2	0.08	0.14	0.02	0.03	0.02	0.03
482.5	0.20	0.14	0.02	0.12	0.04	0.03
520.8	0.09	0.26	0.04	0.16	0.07	0.05
530.9	0.09	1.27	0.08	0.02	0.06	0.02
568.2	0.06	0.10	0.09	0.04	0.07	0.08
647.1	0.10	0.20	0.07	0.10	0.10	0.07
676.4	0.14	0.15	0.04	0.02	0.06	0.03

Before continuing with the results discussion, it is nec essary to remark the following:

a) E.C.P.R.: it is a ten years old radiometer. No change has - been noticed on correction factors for the optical electrical equivalence. The original calibration factor is still being used.

b) N.P.R.L.: it is a four years old radiometer. Standards cali bration factors, given by the manufacturer, are still being used.

Radiometers based on silicon photodiodes have been kept in a dark and dry environment while they were not used for mea-surements.

In order to explain the results shown in Figure I some other sources of error have to be considered. The two thermal radiometers used have a nominal accuracy of 1%. Besides, con-sidering the beam diameter, the corresponding non uniformity uncertainty is $\pm$ 0.3% for E.C.P.R. and $\pm$ 0.4% for N.P.R.L.. Therefore, the results obtained with both radiometers are with-in the total uncertainty interval.

The Q.E.D. 100 is supposed to be accurate in tens of a percent. No other data are available for other error sources, so the results obtained cannot be explained in terms of accura cy. Nevertheless the maximum difference obtained between the flux measured by this radiometer and the average flux is 1%, which is common in commercial radiometers.

With respect to the results obtained with the silicon - photodiodes (Figure 2) only mention the following: the detector A-1 (marked with * in the graphic) shows lower differences for the shorter wavelengths; these differences are very closed to the uncertainty estimated for this detector (2%). The detectors A-3 and PAC-2 (marked with + and # in the graphic) show lower differences for the longer wavelengths of the visible spectrum, being in this region at the same order than the estimated uncer tainty, which is also 2%.

CONCLUSIONS

The main conclusion of this work is that the agreement between the measurements with the different absolute radiome-ters used is $\pm$ 1%.

It could be also concluded that the agreement of differ ent absolute radiometers in measuring radiant flux is better for longer wavelengths than for shortest in the visible spec-trum.

REFERENCES

Geist J 1972 NBS Technical Note 594-1.

Geist J and Blevin W R 1973 Applied Optics 12 2532.

Geist J Zalewski E F Schaefer A R 1980 Applied Optics 19 3795.

Hengstberger F 1977 Metrologia 13 69.

Stock K D 1987 Measurements 5 141.

Zalewski E F and Duda C 1983 Applied Optics 22 2867.

Inst. Phys. Conf. Ser. No. 92
Paper presented at Int. Conf. Optical Radiometry, NPL, London, 12–13 April 1988

159

Calibration of fiber optical power meters at PTB

K D Stock

Physikalisch-Technische Bundesanstalt, Postfach 3345, D-3300 Braunschweig
FRG

ABSTRACT: The calibration of optical power meters at PTB is based on
national primary detector standards. This paper describes the calibration procedure. The achievable uncertainty of power meter calibrations
in PTB at the three optical windows - 850 nm, 1300 nm, and 1550 nm - is
about $\pm$ 0.5 % (1σ uncertainty). The PTB facility for the measurement of the nonlinearity of detectors is also described. All devices
are free air set-ups. The optical set-ups are aligned so as to avoid
the disadvantageous interreflections between detector and back-reflecting surfaces.

1. INTRODUCTION

In the field of fiber optic technology, systems are tested with optical
power meters instead of the volt meters used for conventional communication systems. The demand for traceability of these power meters back to
national standards is clearly increasing. The steps of the PTB radiant
power scale transfer to usual power meters are outlined in the following
simplified table:

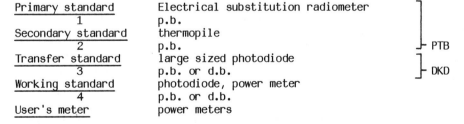

Primary standard	Electrical substitution radiometer	
1	p.b.	
Secondary standard	thermopile	
2	p.b.	PTB
Transfer standard	large sized photodiode	
3	p.b. or d.b.	DKD
Working standard	photodiode, power meter	
4	p.b. or d.b.	
User's meter	power meters	

Table 1.   Transfer steps of the PTB scale of radiant power to usual power
meters. DKD: Deutscher Kalibrierdienst (German Calibration Service). p.b.:
nearly parallel beam. d.b.: divergent beam.

The radiation exits at the end of a fiber in a diverging cone shaped beam.
The angular distribution of radiant power is dependent on the particular
type of fiber. The beam shape is roughly characterized by the half-angle
of the cone or the numerical aperture which ranges from about 0.1 to 0.4.
The fiber beam gives rise to three additional sources of error in measurements; i) the angular dependence of the detector's responsivity, ii) the

speckle pattern on the detector's surface if coherent sources are used, and iii) an additional amount of interreflected radiation increasing the radiation signal. The interreflections are caused by the short distance between the fiber/connector combination and the detector. This distance has to be in the millimeter range in order to guarantee that the cone shaped beam is completely picked up. The amount of interreflected radiation usually varies with wavelength because of the wavelength dependent reflectances of connectors and detectors (Stock 1987). The interreflected radiation is therefore specific to each individual combination of connector and detector. The interreflections are suspected to be the main reasons for the large spread which Gallawa (1986) found when he compared commercially available power meters.

At PTB, the additional sources of error mentioned above are avoided in the first stages of the standard's transfer. The PTB facilities use nearly parallel beams, almost non-coherent sources, and long distances between the detector and the closest back-reflecting surfaces.

At present, PTB calibrates transfer standards for all users; the calibration laboratories of the German Calibration Service (1983), the DKD, are in the process of being set up. The DKD is a system of accredited calibration laboratories authorized to issue calibration certificates which demonstrate <u>authentic traceability</u> to national and international standards (Fay 1988).

2. PTB PRIMARY STANDARDS

PTB optical detector calibrations are based on electrical substitution radiometers (ESR). The laboratory for radiometry in Braunschweig has created different types of primary standards for a variety of power ranges and sources (Stock 1985, Möstl 1988). The primary standards have been subjected to very thorough international intercomparison. The fiber power meters are based on a set of four ESRs which operate in the mW range. The ESRs differ in their type of absorber: two are of the disk type and two are of the cavity type (Bischoff 1968). The four absorbers are coated with black paint (Nextel 3M). Operating in room atmosphere without any window, the disk radiometers have well known absorptances of 0.9734 in the visible range. The cavity radiometers have absorptances of 0.9995 within the spectral range of 250 nm up to 20 μm (Bauer and Bischoff 1971). The uncertainty for the irradiance scale based on this set of ESRs is $\pm$ 0.1 %. The uncertainties given in this paper are 1σ uncertainties with respect to those being taken as a basis in this conference (PTB certificates usually state 2σ uncertainties).

In the radiation field of a standard tungsten lamp of the Osram Wi40 type, the irradiance scale is transferred from the ESRs to a PTB secondary standard. A glass plate cuts off the far IR at a wavelength of about 2.8 μm (Blevin and Brown 1967) to avoid the influence of the lamp's bulb radiation. The room temperature and humidity is kept constant at 23 °C and 50 %, respectively. The secondary standard is a pair of commercially available compensating thermopiles (section 2 and step 2 in table 1). The reproducibility of the thermopile responsivity is $\pm$ 0.09 %.

3. SPECTRAL CALIBRATION OF TRANSFER STANDARDS

Large area Ge photodiodes (5 mm in diameter) are well qualified as standards in the spectral range from 700 nm to 1600 nm (Stock and Möstl 1982,
Stock 1987, 1988). This type of diode is the favoured transfer device
between our national standards and the working standards.

In PTB photodiodes are spectrally calibrated (step 3 in table 1) in two
separate steps - the absolute calibration at one wavelength λ_0 and the
relative calibration $s(\lambda)/s(\lambda_0)$ in the wavelength range of interest. The
spectral responsivity $s(\lambda)$ results from

$$s(\lambda) = s(\lambda_0) \cdot s(\lambda)/s(\lambda_0) \qquad\qquad (1)$$

3.1 Absolute responsivity

The optical set-up for the absolute responsivity $s(\lambda_0)$ measurement is
shown in figure 1. A set of medium-pressure Hg lamps produces a stable
field of radiation without any imaging. The radiation field at a distance
of 1 meter from the source has a sufficiently homogeneous area of about
50×50 mm^2 where the radiation is nearly parallel with a divergence of
about 0.05 rad (half-angle). Filters are used to select a line from the
lamp's spectrum to produce a monochromatic field. The wavelength of $\lambda_0 =$
1014.0 nm is chosen for the calibrations in the fiber optical range. In
the above defined area, the irradiance E of about 0.25 W/m^2 is measured
by the PTB secondary standard.

An aperture of well characterized area A and the photodiode transfer
standard are aligned at the measurement position of 1 meter. The beam
shaped by the aperture must be completely picked up by the diode. The
radiant power received, ϕ, is $\phi = E \cdot A$.

The constant temperature of the lamp housing and the shutter in figure 1
minimises thermal drifts of the thermal detector standards. For Ge diodes

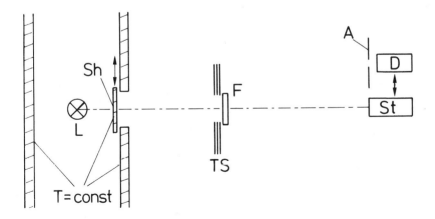

Figure 1. PTB facility for absolute responsivity calibrations $s(\lambda_0)$ of
transfer standards at λ_0. L: Hg lamp; Sh: shutter; TS: triple shield; F:
filter; A: aperture; D: detector; St: standard. Distance from lamp to
standard: 1 meter. Divergence: 0.05 rad. Aperture A: 3 mm in diameter.

of 5 mm in diameter, apertures with a diameter of 3 mm are used. The transfer standard diode in figure 1 receives a radiant power level of typically 2 µW at the wavelength λ_0 = 1014.0 nm. The achievable 1σ uncertainty of the absolute responsivity of a Ge diode is about $\pm$ 0.3 %.

3.2 Relative spectral responsivity

Figure 2 shows the optical set-up for of the relative spectral responsivity $s(\lambda)/s(\lambda_0)$ measurement. The radiation source is a flint-glass double-monochromator and halogen lamp combination that covers the wavelengths from about 400 nm to 2500 nm. The relative spectral responsivity of the detector is measured by a comparison with the secondary standard. A spherical swivel mirror switches alternately between two courses of principle beams; i) the exit slit is imaged to the standard and ii) the aperture A which is irradiated by the exit slit, is imaged to the detector by using a flat mirror and a spherical swivel mirror combination. The secondary standard for relative responsivity measurements is a Hilger & Watts Ft15 thermopile with a small time constant (0.1 s) and high sensitivity (20 V/W). The relative spectral responsivity of this thermopile was found to be constant within the spectral range of 850 nm up to 1900 nm within an uncertainty of about $\pm$ 0.05 %. Its relative spectral responsivity was determined by comparisons with a thermal cavity detector (Bischoff 1964). An Hg-lamp is used to routinely check the wavelength calibration of the monochromator. The cone half-angle of the monochromatic beam irradiating the detector is about 0.05 rad.

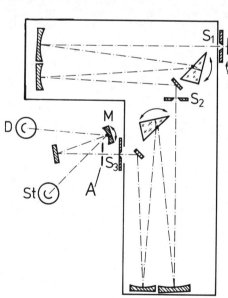

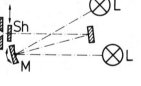

Figure 2. PTB facility for relative responsivity calibrations $s(\lambda)/s(\lambda_0)$ of transfer standards using a flint-glass prism double-monochromator. L: lamp; M: spherical swivel mirror; Sh: shutter; S: slit; A: aperture; D: detector; St: standard.

Figure 3 is a graph of the radiant power of this set-up versus the wavelength. The curve is essentially determined by the spectral radiation function of the lamp, the dispersion of the prisms and the spectral reflectances of the mirrors. The spectral bandwidths shown in figure 4 are determined by the dispersion of the glass prisms. The uncertainties of the relative responsivities are about $\pm$ 0.3 %. The total uncertainty for the spectral responsivity $s(\lambda)$ is about $\pm$ 0.5 % (quadratic addition).

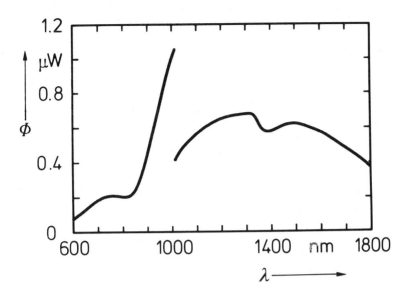

Figure 3.   Output power φ versus wavelengths λ of the monochromator facility. The gap of the curve is caused by a change of the geometrical slit width.

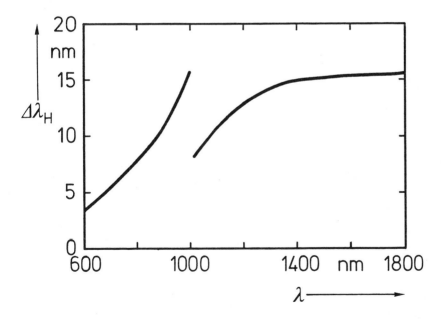

Figure 4.   Spectral half-band width of the   flint-glass double-monochromator. The gap: see figure 3.

4. NONLINEARITY MEASUREMENT

The photodiodes are calibrated at the radiant power of some µW (see section 3 and figure 3). At this level the responsivity of Ge diodes is usually independent of the power level. Fiber optical power meters are used up to the mW range; in this range the detectors should have their linearity checked. At PTB the nonlinearity function of responsivity versus radiant power can be measured with high precision. An automated set-up based on the addition method (Bischoff 1961, Budde 1983) is used for the nonlinearity measurements.

In a one-to-one ratio, a source aperture of 3 mm in diameter, homogeneously irradiated by a halogen lamp system, is imaged onto the center of the detector's active area. Two 100 mm diameter achromatic lenses with a focal length of 1000 mm are used to image the beam onto the detector (figure 5). A system of apertures, located between the lenses and controlled with pneumatically operated shutters, produces the partial beams a and b. The diode is irradiated in a sequence by the beams a, b and a+b, producing the currents I_a, I_b and I_{a+b}. The homogeneity of radiant power on the diode's surface is not affected by this procedure. The nonlinearity for the doubling step of radiant power is

$$f_{1:2}(I_{a+b}) = I_{a+b}/(I_a - I_b) - 1 \qquad (2)$$

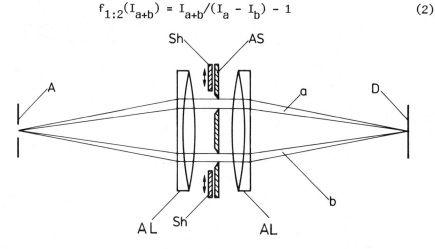

Figure 5.   PTB facility for nonlinearity measurements. Addition method (simplified figure). A: source aperture; Sh: shutter; AS: aperture system; D: detector; AL: achromatic lens.

The simplified aperture system between the lenses shown in figure 5 contains two apertures, a and b, of equal size which are equivalent to the aperture 1a and 1b in figure 6. The real aperture system in figure 6 contains ten additional apertures with gradually increasing sizes as well. In each of these steps the aperture area is increased by a factor of 2 (see figure 5). Eleven subsequent steps allow the radiant power realized to be increased by using a series of sequentially larger apertures. In the final doubling steps, beam a is the largest of the set and beam b is the sum of all the apertures smaller than the aperture for a. The nonlinearity results of all doubling steps referring to equation 2 are combined mathematically. In this way a power range of $1:2^{11}$ is covered which is about

1:2000. The resulting nonlinearity function is

$$f_{NL}(I) = s(I)/s(I_0) - 1 \qquad (3)$$

Nonlinearities of photodiodes may show a very strong spectral dependence (Stock 1986). To investigate spectral nonlinearities, a set of filters can be fitted into the beams. Interference filters of about 25 nm – 30 nm half-band width are used to obtain sufficiently monochromatic radiation and high power levels. At 1300 nm and 1550 nm, power levels of about 1 mW can be reached within a spot 3 mm in diameter.

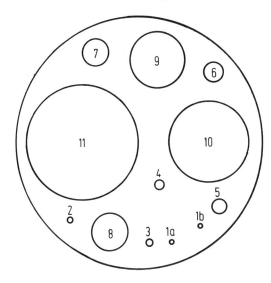

Figure 6. Aperture system of the nonlinearity measurement facility. Apertures: 1a, 1b, 2, ... 11.

The maximum power level in nonlinearity measurements is limited by the lamp's power and the filters used. The lower power limit for one run of mesurements is given by the number of apertures. This limit can be extended by additional sets of measurements at strongly attenuated irradiation of the source aperture. The results can be fitted to cover a multiple of a single 1:2000 power range. The attenuation is realized by defocussing the lamp. Attenuation of the lamp's electrical power is avoided to preserve its relative spectrum. The lowest limit is given by the noise equivalent power level. Cooled Ge photodiodes can be checked for nonlinearity in a power range from 1 µW to 1 mW with an 1σ uncertainty which never exceeds ± 0.005 %.

5. OUTLOOK

Modern commercially available Ge photodiodes are optimized for the 2nd optical window at 1300 nm. The wavelength λ_0 = 1014 nm used for the absolute calibrations in PTB is located outside the optimized wavelength range. Here the reflectance strongly increases with decreasing wavelength and the responsivity is only about 30 % of its spectral maximum. A set-up for absolute calibrations near 1300 nm is in preparation where a laser diode pumped Nd-YLF laser will be used as a single line source.

Bauer G and Bischoff K 1971 Appl. Opt. 10 2639
Bischoff K 1961 Z. Instrumentenkunde 60 113
Bischoff K 1964 Optik 21 521
Bischoff K 1968 Optik 28 183
Blevin W R 1967 Austr. J. Phys. 20 567
Budde W 1983 'Optical Radiation Measurements - Physical Detectors
   of Optical Radiation' (New York, Academic Press) pp 72-80
Fay E 1988 NCSL Newsletter 28 12
Gallowa R L and Yang S 1986 Appl. Opt. 25 1066
German Calibration Service 1983 NCSL Newsletter 1983 23 14
Möstl K 1988 Measurement in press
Stock K D and Möstl K 1982 Proc. 10th Int. Symp. IMEKO TC2 'Photon
   Detectors' (W-Berlin) pp 40-6
Stock K D 1985 Proc. Int. Meet. 'Advances in Absolute Radiometry'
   (Cambridge/Ma) pp42-5
Stock K D 1986 Appl. Opt. 25 830
Stock K D 1987 Proc. 13th Int. Symp. IMEKO TC2 'Photon Detectors'
   (Braunschweig) in press
Stock K D 1988 Appl. Opt. 27 12

Inst. Phys. Conf. Ser. No. 92
Paper presented at Int. Conf. Optical Radiometry, NPL, London, 12–13 April 1988

167

Regeneration of the internal quantum efficiency of Si photodiodes

K D Stock

Physikalisch-Technische Bundesanstalt, Postfach 3345, D-3300 Braunschweig
FRG

ABSTRACT: Long-term stability checks of transfer standard detectors
take up valuable time. A moderate baking procedure and an investigation
into the change in the internal quantum efficiency of Si photodiodes
has been tried out in order to predict stability. Stable diodes which
were observed over years remained stable during and after the baking
procedure. Diodes whose internal quantum efficiency decreased in the
blue and UV spectral range, were regenerated by baking. However, these
diodes subsequently continued to age.

1. INTRODUCTION

Several commercially available and very common types of Si photodiodes
used as transfer standards in the field of radiometry and photometry have
very poor long-term stability. Particularly in the blue and ultraviolet
wavelength range, a strong temporal decrease in the responsivity has been
observed (Gardner and Wilkinson 1985, Stock and Heine 1985, Stock 1987,
Korde and Geist 1987). A more extensive stability test program for Si photodiodes funded by BCR is in progress, with the NPL, INM, IEN, and PTB
taking part. In the investigations described here, a pre-ageing method was
tested. In order to accelerate the ageing process and to achieve a more
stable final state, the diodes were treated by moderate baking. This resulted in the ageing process being reversed instead of accelerated. The
values of the internal quantum efficiency $\eta(\lambda)$ of the diodes approached
those measured at the beginning of the test program.

2. BAKING EXPERIMENT

The responsivity $s(\lambda)$ of Si photodiodes of various manufacturers and types
were monitored in PTB over a period of many years (Stock 1987). The wavelengths λ ranged from about 250 nm to 1200 nm. The active areas of the
diodes are 100 mm² and are encapsulated in housing with quartz windows.
The photodiodes were stored and measured at room temperature. For the
baking experiment, three of the monitored diodes were chosen. These were
placed in a glass box and baked at 80 °C for a period of 41 days. The
responsivity $s(\lambda)$ and the reflectance $\rho(\lambda)$ of the diodes were measured
immediately prior to and after the heat treatment.

The spectral responsivities measured are referred to the PTB's detector-
based absolute radiant power scale (ESRs, Bischoff 1968, Stock 1985, Möstl

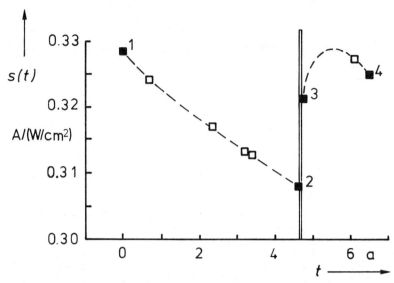

Figure 1. Ageing and regeneration of the responsivity of a EG&G-UV444B photodiode by baking monitored at the wavelengths 546.1 nm over the time t. The black numbered squares mark the points which are related to the curves of figure 2 with the same identification number. The vertical bar marks the heat treatment of 80 °C over the period of 41 days. The dates of measurements are 1: September 81; 2: April 86 just before the baking; 3: June 86 just after the baking; 4: March 88.

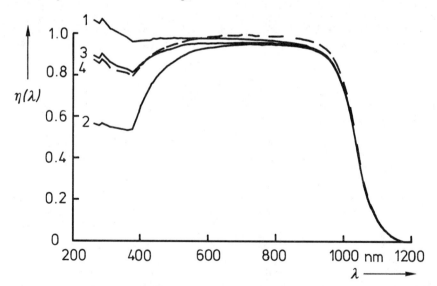

Figure 2. Internal quantum efficiency $\eta(\lambda)$ of a EG&G-UV444B photodiode versus wavelength λ calculated by equation 1 from measured responsivities and reflectances. The dates of the curves are 1: September 81; 2: April 86 just before the baking; 3: July 86 just after the baking; 4: March 88 (dashed line).

1988). The measurements were performed with the PTB calibration equipment (Stock 1988). The sum of the diffuse and specular reflectance $\varrho(\lambda)$ was measured by a spectrophotometer (Varian, Cary UV-VIS) with a modified integrating sphere accessory.

3. EFFECT ON RESPONSIVITY

The left part of the curve in figure 1 shows the temporal decrease in responsivity $s(\lambda)$ of an EG&G-UV444B Si photodiode at one line of the mercury spectrum $\lambda = 546.1$ nm. This diode shows very distinct ageing, and therefore was chosen as an example to illustrate the temporal change. This ageing effect is very typical of this type of diode (Stock 1987).

After the responsivity was monitored over a period of about 4.5 years, the responsivity increased by the baking procedure. The diode continued to regenerate after the heat treatment had been finished. However, first results show that the diode ultimately continues to age. The responsivity of this figure is related to the irradiance E (in W/cm²).

4. EFFECT ON THE INTERNAL QUANTUM EFFICIENCY

The number of incident photons reduced by the reflectance loss $\varrho(\lambda)$ related to the number of output electrons is called the internal quantum efficiency ( ). It is calculated from measured responsivities $s(\lambda)$ and reflectances $\varrho(\lambda)$ by

$$\eta(\lambda) = s(\lambda)/\left[s_{th}(\lambda)\ (1 - \varrho(\lambda))\right] \tag{1}$$

Absorption losses within the diode's window and the antireflection coating are neglected. The theoretical responsivity $s_{th}(\lambda)$ comprises the assumption that one photon generates exactly one electron-hole pair within the semiconductor of the diode. $s_{th}(\lambda)$ is calculable from the wavelength λ and basic constants by

$$s_{th}(\lambda) = \lambda \cdot e/(hc) \tag{2}$$

with the electron charge e, the Planck constant h, and the speed of light c. An ideal loss-free diode has a 100 % internal quantum efficiency.

Since the change in reflectance in all the diodes during the baking did not exceed the measurement uncertainty, we were encouraged to extrapolate the reflectance data to the very beginning of the responsivity monitoring program. Figure 2 shows the temporal change of $\eta(\lambda)$ of the EG&G-UV444B diode used in figure 1. The original curve 1 is close to the 100 % level for wavelengths from 250 nm to about 900 nm. Within five years, $\eta(\lambda)$ dropped by more than 50 % in the blue and UV spectral range. The baking process regenerated the internal quantum efficiency. An ageing of about 3 years was cancelled by this treatment. However, as shown in figure 1, the responsivity continues to age. The responsivity related to irradiance was converted to the responsivity related to radiant power using the 1 cm² value of the active area of the diode.

The older diode type EG&G-UV444A in figure 3, shows a high internal quantum efficiency similar to that of the B-type although the responsivity is lower due to the higher reflection losses. This A-type diode is a more stable device. The baking procedure raised $\eta(\lambda)$ to that measured about 6.5 years before. After finishing the heat treatment $\eta(\lambda)$ returned to the level measured previously.

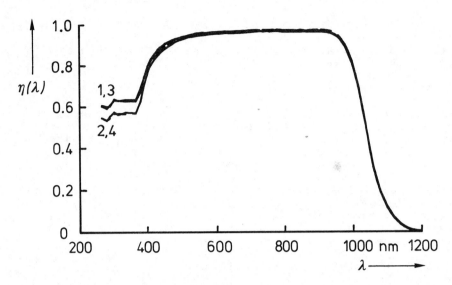

Figure 3.   Internal quantum efficiency $\eta(\lambda)$ of a   EG&G–UV444A photodiode. The   dates   of the   curves are   1: November   79; 2:   April 86   just before baking; 3: July 86 just after baking; 4: March 88 (dashed line).

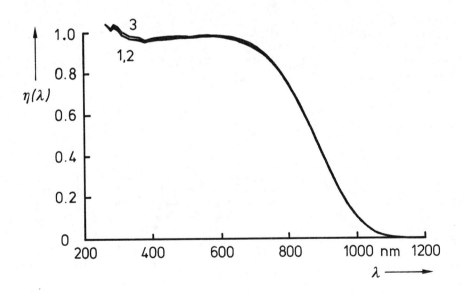

Figure 4.   Internal quantum   efficiency $\eta(\lambda)$ of a   Hamamatsu S1227–1010BQ photodiode. The dates of the curves   are 1:   August 85;   2: April   86 just before the baking; 3: July 86 just after the baking.

The Hamamatsu S1227-1010BQ photodiode has an almost 100 % internal quantum
efficiency below about 600 nm (see figure 4). Neither the storage nor the
baking had much effect on the internal quantum efficiency. The changes did
not exceed the measurement uncertainty. Unfortunately, up to the deadline
for this paper, a repeat of the responsivity measurements on this diode
have not been concluded.

5. CONCLUSION

The three diodes dicussed in section 4 vary considerably in their
stability. The more unstable they are, the more the baking regenerates
internal quantum efficiency. From the results obtained with these three
samples, it would seem that a heat treatment like this is a helpful tool
in dealing with the problematic question of a long-term stability
prognosis for photodiodes. Of course, this experiment may produce the
wrong answers for very freshly manufactured and non-aged diodes.

It is remarkable that there is a similarity between the spectral pattern
of ageing and the regeneration if we neglect the sign of change. These
changes in the internal quantum efficiencies are apparently caused by the
same internal processes with opposite signs. Referring to argument presented in a previous paper (Stock 1987), the temporal migration of sodium
ions within the SiO_2 antireflection coating towards a low temperature
equilibrium of local concentration is reversed by baking towards a
different high temperature equilibrium. After the baking process has been
finished the sodium ions again follow their original tendency. The ion
concentration in the vicinity of the $Si-SiO_2$ interface strongly influ-
ences the recombination rate of charge carriers in the Si wafer. Because
of the wavelength-dependent penetration depths of photons in the wafer,
the ion migration predominantly affects the charge carriers which are
generated by the shorter wavelengths.

The specifying of commercial devices in this work does not imply endorsement by PTB or that these devices are the best available for a particular
application.

The author wishes to thank K. Möstl for suggestions and helpful discussions and H. Heine for carrying out the responsivity measurements.

Bischoff K 1968 Optik 28 183
Gardner J L and Wilson F J 1985 Appl. Opt. 24 1531
Korde R and Geist J 1987 Appl. Opt. 26 5284
Möstl K 1988 Measurement in press
Stock K D and Heine R 1985 Optik 71 137
Stock K D 1985 Proc. Adv. in Absolute Radiometry (Cambridge/Ma) pp 42-5
Stock K D Measurement 1987 5 141
Stock K D 1988 seperate paper in this proceedings

Inst. Phys. Conf. Ser. No. 92
Paper presented at Int. Conf. Optical Radiometry, NPL, London, 12–13 April 1988

Reduction of the responsivity of the ROSAT focal plane x-ray detector (PSPC) to ultraviolet radiation and absolute calibration of its UV-quantum efficiency

K.H. Stephan[1], U. Briel[1], and H. Kaase [2]

[1]Max-Planck Insitut für Physik und Astrophysik, Institut für extraterrestrische Physik, Garching, F.R.G.
[2]Institut für Lichttechnik der Technischen Universität Berlin,

Abstract .Two redundant multiwire proportional counters (PSPC's) are serving as imaging detectors in the focal plane instrumentation of the German X-ray astronomy satellite ROSAT (Röntgensatellit), Since commercially not available the detectors are equipped with a self developed radiation entrance window type, The window has to meet demands for high transmittance in the soft X-ray spectral range ($0,6 < \lambda < 10$ nm) while it must be opaque for radiation in the ultraviolet (UV) spectral range ($100 < \lambda < 250$ nm) in order to prevent contamination by the very intense geocoronal UV-lines and by UV-stars, The window is coated with a protective layer of a polycarbonate, leading to a transmittance of
$$\tau(\lambda) < 10\text{-}6 \quad \text{at 175 nm}$$
which provides a sufficient suppression of of UV- photons from the counter, Measurements of the performance data of the radiation entrance windows as well as irradiations tests on the complete PSPC show that the requirements for a quantum efficiency
$$(\lambda) < 10^{-10} \text{ electrons/photons}$$
in the UV-spectral range ($\lambda < 250$ nm) could be met,
The experimental method of determining the UV-quantum efficiency of the PSPC as well as the measurements of the window transmittance are described and the results are presented,

1. Introduction :

The focal plane assembly (FI) of the ROSAT [1] telescope comprises two redundant imaging PSPC's. The instrumentation was designed and built at the Institut für Extraterrestrische Physik of the Max Planck Institut für Physik und Astrophysik at Garching / FRG. The telesccope system consists of four nested Wolter I type mirrors, the outermost of which has an aperture of 83 cm.
Fig. 1 shows a cross-section of the telescope and the focal plane detectors which are mounted on a carousel platform. At present final performance tests are being executed with the flight model FI. The ROSAT observatory is scheduled to be launched into a 57 degree inclination orbit at 475 km height on board a Delta rocket early in 1990.

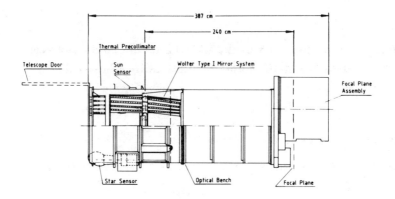

Fig. 1   Scheme of the ROSAT-telescope system and focal plane assembly

2. Objective of the work

The manufacturing procedure and performances of the radiation entrance window of the PSPC are described in detail elsewhere [2]. Among a number of demanding requirements the window must feature relatively high transmittance in the soft X-ray spectral range;

$$\tau(\lambda) \geqslant 0.50 \text{ at energies of 1 and 0.28 keV,}$$

while the transmittance in the UV-spectral range $\lambda \leqslant 250$ nm must not exceed values as high as

$$\tau(\lambda) = 10^{-6}$$

The opacity to radiation in the UV-region becomes necessary in order to avoid an increase of the background level of the PSPC caused by UV-photons. In the interplanetary space intense UV-radiation is emitted from the geocorona, the earth's albedo and O, B, and A stars. The UV solar radiation scattered in the geocorona dominates by orders of magnitude, where essentially, atomic lines of hydrogen, helium and oyxgen (HI, HeI, HeII, OI, OII) and molecular lines of nitrogen (N_2) are concerned. Based on stellar fluxes as given in [3] and rocket measurements of the airglow and aurora [4,5] it can be estimated that the irradiances in the $\lambda \leqslant 250$ nm wavelength range do not exceed

$$E(\lambda) \geqslant 6 \times 10^{-11} \text{ W/cm}^2$$

in the telescope's incidence plane during the mission of ROSAT.
It is the objective of this paper to describe the protective coating which is applied to the radiation entrance window, and shows how the reduction of UV-radiation was found to prove satisfactory. Before concentrating on this task the PSPC is briefly explained.

3. The PSPC

The PSPC is a multiwire proportional counter with a CHARPAK chamber geometry as shown in figure 2. The detector is divided in two counters: anode A1 together with cathodes K1 and K2 represent the position and energy sensing part; Anode A2 with cathode K3 is an anticoincidence counter for background rejection. The electrodes are wire grids, epoxied and soldered to glass ceramic frames. The grid system is mounted to a bottom plate (see figure 3) and accomodated in a counter housing. The detector

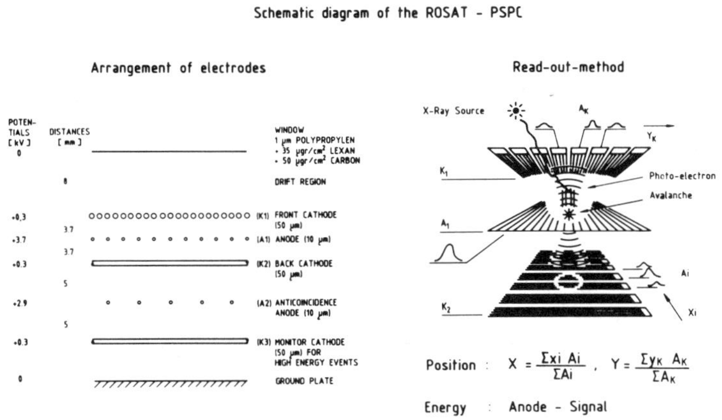

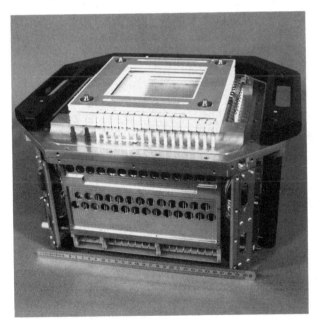

Fig. 2   Principle of operation of the PSPC

Fig. 3   View of the PSPC with cover removed

is filled with a gas mixture (65% Ar, 20% Xe, 15% CH4$_4$) at a pressure of 1.45 bar. The gas is density regulated and refreshed with a flow rate of 5 cm^3/min. The top of the housing holds a circular window of 80mm in diameter which is described in the next chapter. The PSPC is designed for an energy range from 0.1 to 2.0 keV X-rays.
The energy resolution is

$$\Delta E_{FWHM}/E = 0.4 \times E^{-1/2} \qquad \text{(E in keV)}$$

The mean position resolution is

$$\Delta X = 250 \text{ and } 400 \text{ } \mu m \qquad \text{(FWHM)}$$

for 0.93 and 0.28 keV X-rays respectively. A detailed description of the PSPC, the corresponding electronics and the performance of the PSPC is given in [6,7,8].

4. The radiation entrance window :

The window has an effective diameter of 80 mm and consists of a 1μm thick polypropylene (Novolen 1300)-foil as a basic material which is coated with a conductive graphite suspension on the inside and a polycarbonate (Lexan)-deposit on the outside. It is supported by an especially designed supporting grid system which comprises a rigid circular mainframe with a concentric ring of 28 mm aperture and 8 radial struts combined with two grids. The course mesh is built by 100 μm gold plated tungsten wires with 2 mm spacings; the fine grid has 25 μm gold plated tungsten wires 0.4 mm spaced. Thus a transmission factor of 72 percent due to geometrical obscurations is achieved. The window foil which is glued on an aluminium ring is fixed by an adjusting flange and tightened by an O-ring. Figure 4 shows an exploded view of the window assembly.

Fig. 4   Exploded view of the radiation entrance window.

The window is adapted to the front end of the PSPC-housing and must withstand an operating pressure of 1500 mbars in the counter tube during the mission of the satellite which is planned to last for 1.5 years. During this time the gas leakage must not exceed

$$Q = 5 \times 10^{-3} \text{ mbarl/s}$$

5. Suppression of the UV-responsivity of the PSPC

It is well known that the molecule " Bisphenol A " which is a part of the monomeric repetition unit of polycarbonate [$C_{16}H_{14}O_3$]n (trade names : Lexan, Makrolon) has outstanding absorption properties in the UV-range. Therefore the front surface of the window was covered with a uniform coating of polycarbonate by centrifugation of soluted Lexan. As described later we chose a thickness of this layer corresponding to a mass density of 35 µg/cm² which provides a sufficient suppression of UV-photons.

6. Measurements of the performance

The measurements of the spectral transmittance of window foils as well as the determination of the responsivity of the PSPC in the UV-spectral range were performed in a UV-facility of the Physikalisch Technische Bundes-anstalt (PTB) at Braunschweig / FRG. This experimental set-up mainly comprises a 140 MeV- synchroton and a wall stabilized Argon arc [9,10] as radiation sources. Both sources are used as radiation standards, and they are combined with an 1 m Mc Pherson normal incidence vacuum-monochromator. The optical arrangement is schematically shown in figure 5. The electron orbit of the synchrotron (2) and the Ar-arc

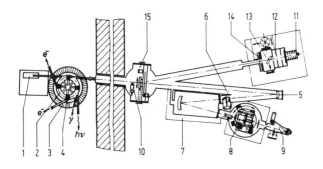

Fig. 5   UV – facility of the PTB

source (11) are imaged onto the entrance slit of the monochromator (7) by
using a spherical mirror (5). For spectral transmission measurements the
test objects are situated either in the vacuum chamber (15) or in the test
chamber (8) with a goniometer. The spectral transmittance of different
coatings are determined using a 3 mm thick LiF-substrate; parameter for
such a measurement was the thickness of the coated layer. To reduce the
influence of stray radiation special interference filters in position (6)
and solar blind detectors are used; with the synchrotron as the radiation
source we succeeded in determining spectral transmittances as low as 10^{-7}.
For the calibration of the PSPC in the spectral range from 120 nm to
300 nm, the detector was mounted behind the goniometer chamber at position
(9). The Ar-arc source was operated as the radiation source and the
measured photocurrent compared with the photocurrent of a calibrated
photodiode used as an UV transfer standard detector [11]. The calibration
of the photodiode (standard detector) was also carried out with this
facility, using the Ar-arc as a standard source of radiation intensity. From
this and from a measurement of the spectral transmittance of the optical
arrangement with an additional monochromator (9), the monochromatic flux
reaching the detector could be evaluated. The spectral photon irradiance
on the PSPC was

$$E_P (\lambda) = [\delta E_P (\lambda)/\delta\lambda]\Delta\lambda \ngtr 10^{10} \text{ photons/cm}^2\text{s}$$

The bandwidth of the monochromator was 1.7 nm.

7. Results :

The spectral **absorptance characteristics** in the UV-range of a number of
foil specimen and coatings was studied. After selection of the suited
materials, we investigated the dependance of the transmission on the
thickness and performed **irradiation tests** in order to see whether or not a
deterioration by radiation damages could happen. These results led to
the preparation of the window foils. The local distribution of the spectral
transmittance was scanned across the effective window area.
Finally we exposed a **PSPC** to high UV-radiant fluxes in order to measure
the **responsitvity**.
The results are given in the following sections and corresponding figures.

7.1   UV-transmission of different foil specimen (Fig. 6).

[1] PP-foil 1µm thick
[2] PP-foil 1µm thick coated with 20µg/cm² PC (Lexan, Makrolon),
[3] PP-foil 1µm thick coated with 50µg/cm² C (Graphite),
[4] PP-foil 1µm thick coated with 20µg/cm² Lexan and 50 µg/cm² graphite,
where is denoted :
PP: Polypropylene, PC: Polycarbonate, G: Graphite (Acheson suspension).

The absorption edge towards short wavelengths are found at 230 and 165 nm
for polycarbonate and polypropylene respectively. PP features a monoton-
ously falling transmission while PC shows a further maximum at 175 nm and
a minimum at 195 nm.
A PP-foil coated with a graphite suspension of the given thickness shows
a rest transmittance of $\approx 5 \times 10^{-3}$ peaked at 180 nm and has a steep cut off

towards short wavelengths at 170 nm and a smoother falling tendency
towards long wavelengths reaching a transmittance of
$$\tau(\lambda) \simeq 10^{-5} \text{ at 200 nm.}$$

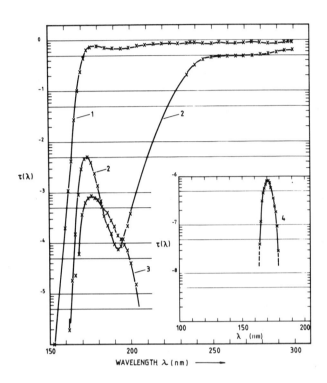

Fig. 6   UV-transmittance of different foil specimen.

7.2 UV-transmission of Lexan (Fig. 7).

The results of measurements on different Lexan films having mass densities
from 10 to 40 μg/cm² are shown. The films were deposited on a LiF-substrate which did'nt affect the results of interest. The transmission of pure
Lexan features a rest maximum at λ≈ 160 nm. For a PP-foil (broken line)
which is coated with Lexan a maximum results for the total transmittance
at λ≈ 170 nm from a multiplication of the corresponding individual transmittances.

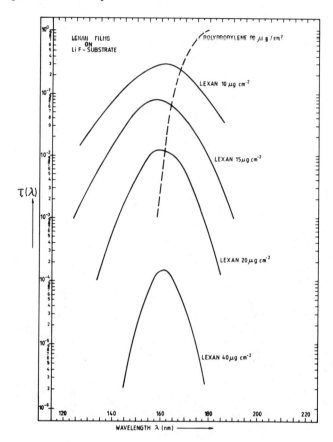

Fig. 7   UV-transmittance of Lexan.

7.3   UV-transmittance of Lexan coated PP-foils (Fig. 8).

The relationship between transmittance and thickness of the Lexan deposit
on the PP-foil was determined at three wavelengths : 165, 170 and 175 nm.
For a Lexan-film having a mass density of 45 $\mu g/cm^2$ an attenuation
of 8 orders of magnitude was measured at the given wavelengths, which is
the detection limit of our method. The transmittance as a function of the
film thickness can be expressed by a linear function in a semi-logarithmic
scale.

7.4   Spectral transmittance of a PSPC radiation entrance window (Fig. 9).

The diagram shows the spectral transmittance in the overall wavelength
range of interest 250 ⩾ λ ⩾ 0.6 nm for a complete PSPC-window including the
supporting grid. In the soft X-ray range at energies of 1.5 and 0.28 keV
the transmittance was found to be

$$\tau(\lambda) \geqslant 0.50,$$

while the UV-transmittance (insert) could be held as low as

$$\tau(\lambda) \leqslant 10^{-6}$$

The local uniformity of the transmittances across the window is better
than ± 10 percent.

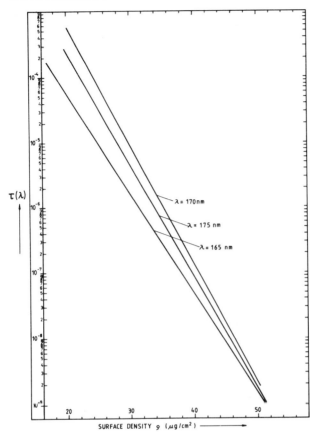

Fig. 8   UV-transmittance of Lexan as a function of the film thickness.

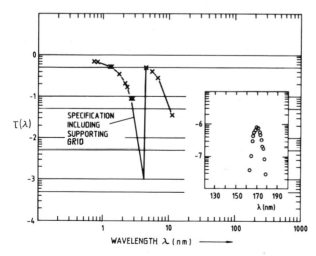

Fig. 9   Spectral transmittance of a PSPC radiation entrance window in the
          soft X-ray and UV-range.

The measurements in the soft X-ray region were carried out in a vacuum facility of the Max Planck Institut at Garching, which comprises an X-ray source emitting characteristic line radiation connected to a 2m-McPherson grazing incidence spectrograph and a specimen chamber.

7.5   Irradiation tests of different foils.

The properties of plastic materials can be changed by irradiation. Polyolefines are considered to be essentially sensitive to radiation. To give an example : photons at energies corresponding to 342 nm and 290 nm are especially affecting the C-C and C-H bindings respectively. Pure PP-foils and Lexan coated PP-foils were exposed to synchroton radiation in order to determine the critical radiation doses which are causing significant deteriorations of the window material.
The monochromatic flux density incident on the foil was in the order of
$$E \simeq 10^{-7} \text{ W/cm}^2$$
in the spectral region below 200 nm having a bandwidth of 180 nm.
The spectral transmittance of the foils was measured before and after the irradiation. After exposure times of 5 hours corresponding to a radiant dose of
$$H = 2 \times 10^{-3} \text{ J/cm}^2$$
the following effects were observed when referring the transmittance of the irradiated to the non-irradiated specimen. :
- For PP-foils the curve shows a minimum at a wavelength of $\lambda \simeq 195$ nm where the quotient drops to a value of 0.5.
- For Lexan coated PP-foils having UV-transmittances which are at least two orders of magnitude lower a similar deterioration is found.

In both cases the relative changes of the transmittances caused by the irradiation are found to depend on the foil thickness; however seem to be independant from the kind of the polycarbonate coating.

7.6   The absolute responsivity of the PSPC in the UV-range. (Fig. 10).

The absolute responsitivity of the PSPC in the UV was derived from the countrate in the PSPC caused by exposure to the radiation of the 100 Ampere cascaded Argon arc. Measurements were performed at the optically thick lines 121.6 (HI), 149.2 (NI), 165.7 (CI), 174.2 (NI), 193.1 (CI), and 247.9 (CI). The quantum efficiency is defined as the ratio of registered PSPC-counts to the incoming number of photons. As the absolutely known incident photon flux was in the order of
$$E_p \geqslant 10^{10} \text{ photons/s}$$
exposure times of 1000 s were applied. The result is that the upper limits of the UV responsivity of the PSPC is determined to be
$$S(\lambda) \leqslant 3 \times 10^{-11} \text{ counts/photon.}$$

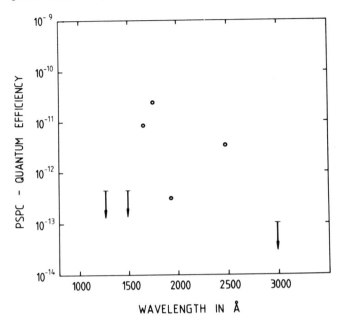

Fig. 10   Absolute quantum efficiency of the PSPC in the UV- range

8. Conclusions

The following conclusions from the previous results, which were enabling us
to design the composition of the window for the flight model PSPC, can be
made :

- **Protective polycarbonate layer.** The investigated materials Lexan (General
  Electric Plastics) and Makrolon (BAYER) show the same spectral
  transmitance in the UV region within the uncertainties of the measuring
  method. We decided on Lexan because it has been successfully used
  for space applications in the past and the selected type 303 is
  especially stabilized against ultraviolet radiation. We have been able to
  make reproducibly uniform layers on the polypropylene foils, which are
  adherent and stable. For a required transmittance of

$$\tau(\lambda) \leqslant 10^{-6} \text{ at 175 nm}$$

  the mass density of the Lexan deposit must be at least as high as

$$\rho = 35 \ \mu g/cm^2$$

- **Conductive layer.** The graphite coating is originally introduced to make
  the window conductive on the inside in order to prevent static charge.
  However it is also important to make the window opaque in the wavelength
  region longward 180 nm. In the counter the element carbon has the lowest
  working function, namely 4.36 eV, among the other used materials. Thus

photons having energies less than 4.36 eV corresponding to 284 nm are considered no longer to deliberate photoelectrons in the counter tube. For a required transmittance of
$$\tau(\lambda) \leqslant 10^{-4} \text{ at 200 nm}$$
a mass density of the Acheson graphite layer of
$$\rho \geqslant 50 \ \mu g/cm^2 \quad \text{is necessary}$$

-**Basic material**. We decided on a polypropylene which was especially made by the BaSF AG/ Ludwigshafen, who doped the material with a high molecular stabilizer with respect to the UV-irradiation levels during the orbital mission of the windows. As far as we know the commercially available PP-foils are stabilized by dopants that have low molecular weights and are considered not to be sufficiently stable under longterm vacuum conditions.

-**Irradiation tests**. UV-irradiances of unprotected as well as protected PP-foils lead to a wavelength dependant decrease of the transmittance. Referring to former tests [12] fractions of irradiated pure PP- foils occurred at radiation dose rates (exposure) as high as
$$H = [\delta H/\delta\lambda]\Delta\lambda \simeq 10^{-1} \ J/cm^2$$

The integral radiation load on the PSPC windows during the mission time can be estimated not to exceed values as high as
$$H \simeq 3x10^{-3} \ J/cm^2.$$
Considering the fact that the Lexan deposit reduces the flux by several orders of magnitude which is entering the PP-foil we conclude that there is no evidence for destruction of the PSPC window by geocoronal radiation. The Lexan layers have proved to be sufficiently resistant and stable.

- **PSPC window**. The required suppression of photons in the UV-spectral range for the PSPC- window can be estimated from irradiances in the telescope's incidence plane by O and B-stars. A number of them shows spectral intensity distributions which are peaked at at $\simeq$ 175 nm, where the window features a maximum in the transmittance curve. From calculations for one of the most intense UV-stars, α Lyr (Vega), results that the responsivity of the PSPC which was measured to be

$$S(\lambda) \leqslant 3x10^{-11} \text{ counts/photon} \quad \text{at 175 nm}$$

is considered to be sufficiently low in order to avoid additional background by UV-photons.

- In summarizing the previous results we designed the composition of the PSPC-window foils as follows.

Material	Trade name	Formula	Mass density $\mu g/cm^2$	Task
		Supporting grid		
Polycarbonate	Lexan 303	$[C_{16}H_{14}O_3]n$	35	UV-Suppression
Polypropylene	Novolen 1300	$[C_3H_6]n$	90	Basic material
Graphite	DAG 154	C-Suspension	60	Conductivity.
		Counter volume		

9. Acknowledgements

We wish to thank Dr. H. Bräuninger for his supporting of the project by critical discussions and proposals. The work of Mrs. E. Künneth who prepared the window foils is greatly acknowledged as well as the contribution of Mr. G. Eckmann who carried out the performance measurements. Finally we wish to express our thanks to Dr. Trier at the PTB for his cooperation in the operation of the synchroton and to Mr. B. Nawo for caring of the experimental set-up.

10. Figure captions.

Fig. 1 : Scheme of the ROSAT-telescope system and focal plane assembly.

Fig. 2 : Principle of operation of the PSPC.

Fig. 3 : View of the PSPC with cover removed.

Fig. 4 : Exploded view of the radiation entrance window.

Fig. 5 : UV-facility of the PTB with specimen chamber for the determination of spectral transmittances of foils.

Fig. 6 : UV-transmittance of different foil specimen.

Fig. 7 : UV-transmittance of Lexan

Fig. 8 : UV-transmittance of Lexan as a function of the film thickness.

Fig. 9 : Spectral transmittance of a PSPC radiation entrance window in the soft X-ray and UV-region.

Fig. 10 : Quantum efficiency of the PSPC in the UV-range.

11.　References

1.　Trümper, J., Adv. Space Research, Vol. 2, No. 4, (1983) 241.
2.　Stephan, K-H., Bräuninger, H., Kaase, H., Maier, H. J., Schöne, W., Wilski, H., SPIE, Vol. 733, (1986) 379.
3.　ESA, SR-28, Supplement to the Ultraviolet Bright Star Spectrophotometric Catalogue, October 1978.
4.　Huffman, R. E., Leblanc, F. J., Larabee, J. C., Paulsen, D. E., J. Geophys. Res. 85, (1980) 2201.
5.　Martin V. Zombeck, Handbook of Space Astronomy and Astrophysics, Cambridge United Press, 1982.
6.　Briel, U. G., and Pfeffermann, E., NIM in Phys. Res., A 242, (1986) 376.
7.　Metzner, G., NIM in Phys. Res., A 242, (1986) 493.
8.　Pfeffermann, E., SPIE, Vol. 733, (1986) 519.
9.　Kaase, H., Stephan, K-H., Appl. Op. 18 (1979) 2275.
10.　Kaase, H., Optik 59 (1981) 1.
11.　Kaase, H., Stephan, K-H., Burton, W. M., Hatter, A. T., Ridgeley, A., Canfield, L. R., and Madden, R. P., Appl. Opt. 19 (1980)2529.
12.　Hovestadt, D., Laeverenz, P., Internal Report / MPE (1973).

Inst. Phys. Conf. Ser. No. 92
Paper presented at Int. Conf. Optical Radiometry, NPL, London, 12–13 April 1988

187

A determination of the thermodynamic temperature of the freezing point of silver by infrared radiation thermometry

J Fischer and H J Jung

Physikalisch-Technische Bundesanstalt, Institut Berlin, Abbestr. 2-12, D-1000 Berlin 10, FRG

ABSTRACT: The thermodynamic temperature of the freezing point of silver is of great importance in the proposed temperature scale (ITS-90), since this temperature fixed point will be the junction between the platinum resistance based subscale and the radiation thermometer based subscale. The determination presented was performed using the dc infrared pyrometer of PTB employing the radiation of a blackbody at the aluminium point as a radiometric reference. The result of this study is t_{Ag} = 961.734 °C ± 0.018 °C (68.3 % confidence level).

1. INTRODUCTION

The thermodynamic temperature of the freezing point of silver is of great importance in the proposed temperature scale ITS-90, since this temperature fixed point will be the junction between the platinum resistance thermometer based subscale and the pyrometric subscale.

Up to now a confirmed set of values for the thermodynamic temperature of the silver point is not available : Only the recent pyrometric determination of Jones and Tapping (1987) based on a reference temperature of 660 °C (Al-point) has yielded a value of satisfactory accuracy (961.744 °C ± 0.060 °C, 99 % confidence level). Based on a reference of 630 °C (Sb-point), but less accurate, are the earlier pyrometric experiments of Bonhoure (1975), recalculated by Tischler and Jimenez Rebagliati (1985) (961.73 °C ± 0.13 °C) and of Andrews and Gu Chuanxin (1984), recalculated by Hudson et al. (1987) (961.82 °C ± 0.10 °C). Further results came from the noise thermometry of Crovini and Actis (1978) (962.00 °C ± 0.34 °C) without requiring a high-temperature reference point.

2. RESULTS

Our determination of the silver point was performed using the dc infrared pyrometer described by Jung (1986) employing two different photovoltaic silicon detectors (Hamamatsu S1336-5BK and Oriel 7180). The pyrometer periodically compared the spectral radiance of a blackbody the cavity of which was submerged in freezing aluminium and another blackbody of the same design the cavity of which was submerged in freezing silver. The purity of both the metals accounts to 99.9999 %. The aluminium sample stemmed of the same batch as the one used for the Al-point determination (Jung 1986). Together with the relative spectral response of the detectors and the reflectance of the mirrors, the employed interference filter (halfwidth 40.7 nm) sets the effective wavelength for this determination to 974 nm and 973 nm, respectively.

The thermodynamic temperature of the freezing point of silver was averaged from about 35 data points per run. Each run included a simultaneous pair of freezes of silver and aluminium, 46 runs have been carried out. The temperature for the Al-point was chosen 660.302 °C according to Jung (1986). The result for the silver sample of 6 n purity is*

$$t_{Ag} = 961.734 \, °C \pm 0.018 \, °C$$

at the 68.3 % confidence level. The uncertainty contributions are listed in table 1. The quoted value for the silver point leads to a deviation $T-T_{68}$ of -196 mK at that temperature.

Table 1   Uncertainty contributions to t_{Ag}

Source	Uncertainty, (1σ) level
Thermodynamic temperature of Al, propagated error	16.3 mK
Relative spectral response of photodiodes	2.2 mK
Non-linearity of photodiodes	3.4 mK
Resistance ratio of feedback resistors used in current to voltage converter	4.0 mK
Spectral transmission of interference filter	4.4 mK
Difference in size of source effects	1.0 mK
Impurity of silver sample	3.0 mK
Random uncertainty	1.0 mK
Addition in quadrature	18.1 mK

Our result is 10 mK lower than that of Jones and Tapping (1987). This is an insignificant difference, because their standard deviation accounts 20 mK even if the uncertainty contribution due to t_{Al} is assumed to be nought.

We aime to extend this work to the gold point in the near future.

REFERENCES

Andrews J W and Gu Chuanxin 1984 *BIPM Com. Cons. Thermométrie* **15** Document CCT/84-39
Bonhoure J 1975 *Metrologia* **11** 141
Crovini L and Actis A 1978 *Metrologia* **14** 69
Hudson R P, Durieux M, Rusby R L, Soulen R J and Swenson C A 1987 *BIPM Com. Cons. Thermométrie* Document CCT/87-9
Jones T P and Tapping J 1987 *BIPM Com. Cons. Thermométrie* Document CCT/87-15
Jung H J 1986 *Metrologia* **23** 19
Tischler M and Jimenez Rebagliati M 1985 *Metrologia* **21** 93

---

* For a silver sample of 5 n purity we found a value 6 mK below the above stated t_{Ag}.

Inst. Phys. Conf. Ser. No. 92
Paper presented at Int. Conf. Optical Radiometry, NPL, London, 12–13 April 1988

The establishment of a new detector based spectral power scale

V E Anderson and N P Fox

Division of Quantum Metrology, National Physical Laboratory
Teddington, Middlesex TW11 0LW, UK

ABSTRACT: The present NPL spectral power scale based on black body radiators (400 nm - 800 nm) has an uncertainty of at best 1%. NPL aims to redetermine the spectral power scale to an uncertainty of 0.3% or better by taking a detector based approach.

1. INTRODUCTION

In most national standards laboratories the scale of spectral irradiance has been realised by a Planckian, black body radiator and is maintained in groups of standard lamps. The realisation of such scales can be very difficult. This was highlighted in 1974 when an international intercomparison of spectral irradiance measurements of tungsten halogen lamps organised by the CIE showed large discrepancies between the participants (Suzuki and Ooba 1976). The results are shown in Fig.1.

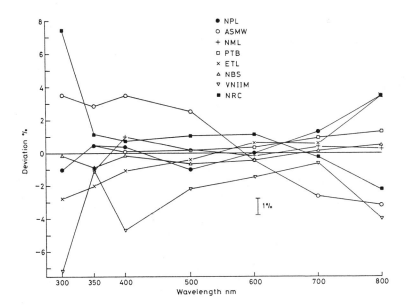

Figure 1. Comparison of spectral irradiance scales maintained at eight national standards laboratories.

The scale realised at NPL (between 400 nm and 800 nm) has an uncertainty of at best 1%. This uncertainty is largely due to problems associated with the use of black bodies, in particular the measurement of temperature and its uniformity. The maintenance of the scale is much easier as the tungsten lamps used as transfer standards are stable and reproducible (to 0.1% or better).

The increasing demand for high accuracy spectral power measurements has lead to a need to improve the basis of the scale. In recent years NPL has developed an absolute radiometric scale based upon a cryogenic radiometer with an uncertainty of 0.005% (Martin *et al* 1985). This detector based radiometry also formed the basis of a new realisation of the SI base unit of luminous intensity, the candela (Goodman and Key 1985), and an intercomparison of the spectral irradiance scale realised by the electron storage ring, BESSY (Fox *et al* 1986). It was therefore appropriate to use these techniques to form the basis of a new realisation of the spectral power scale.

2.   ESTABLISHMENT OF A NEW SCALE

The target for this realistion is to acheive an uncertainty of 0.3% or better and extend the scale to 1000 nm. The method chosen was to determine the absolute spectral irradiance produced by of a coiled tungsten filament lamp at a precisely measured distance using two forms of filter radiometry:

A   Broadband (20 nm) filter radiometry using a set of 13 interference filters which were chosen for their high stability, uniformity, and low stray light characteristics. This approach has already been successful at NPL at two wavelengths (Fox *et al* 1986).

B   Narrowband (1 nm) filter radiometry using a double monochromator at a number of fixed wavelength points in place of the filter. This approach is the more flexible allowing greater freedom to choose the wavelengths.

It is important to know the relative spectral transmission profiles which take a very different form in the two cases. Typical profiles can be seen in Fig.2.

3.   TRANSMISSION PROFILE OF FILTER RADIOMETER

For both narrow and broadband filters the transmission profiles can be accurately determined using an automated dye laser. The measurement techniques used are similar to those first described by Geist et al (1975) and later improved by Schaefer and Eckerle (1984). The dye laser facility is outlined in figure 3. An intensity stabilised dye laser can be used to measure the relative spectral responsivity of the filter radiometer with respect to a spatialy flat (or black) detector. An absolute value can be obtained by making a measurement at a single wavelength (close to the peak) with a transfer silicon photodiode calibrated against the NPL cryogenic absolute radiometer. The wavelength of the dye laser is automatically determined by a wavemeter.

The characterised filter radiometers can then be used to determine the spectral irradiance of the standard lamp at the selected wavelengths. The relatively wide bandpass of the interference filters requires a correction

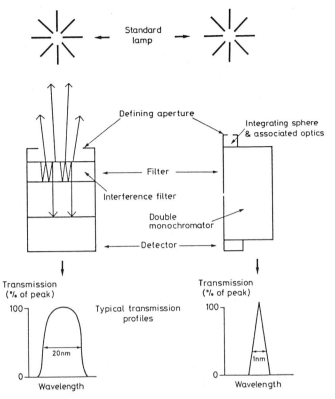

Figure 2. Two types of filter radiometer to be used.

to be applied for the lamps quasi-Planckian power distribution to allow for the difference in spectral emission over the bandpass. This is not necessary in the case of the monochromator because over the narrower bandpass spectrally flat irradiance can be assumed without introducing significant error. The monochromator approach is therefore the more aesthetically correct one. This does however present its own difficulties. The radiation from lamp and laser have very different properties. eg. polaristion, spatial uniformity and coherence. To ensure that the monochromator views and treats both types of radiation in the same manner an integrating sphere is coupled to the monochromator entrance slit. This sphere will act as a secondary source of radiation which can be irradiated by either lamp or laser. The sphere removes problems due to spatial uniformity and polarisation but not coherence.

Because of the lasers coherent properties the output from a laser irradiated integrating sphere is not a uniform field, but a speckle pattern. The problem can be removed by passing the light through a spinning diffuser, vibrating mirror or some other means of changing the optical path. This produces a speckle pattern which is rapidly changing, giving the same effect as a uniform field (radiation from a thermal light source is in fact a speckle pattern changing at 10^{15} times a second).

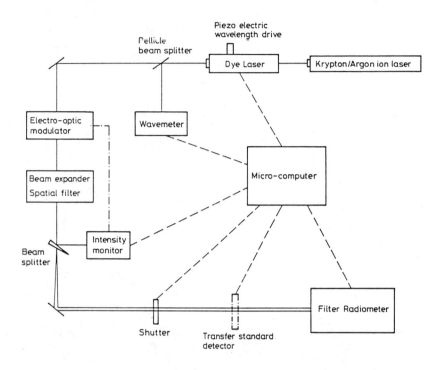

Figure 3. Schematic representation of the dye laser characterisation facility to be used to calibrate the filter radiometers.

These two approaches can be intercompared, giving a check on any errors in either method. There are plans to extend the range of this detector based method into the ultraviolet. Frequency doubling techniques would be employed to perform the laser characterisation of the filters.

4.   REFERENCES

Fox N P, Key P J, Riehle F and Wende B 1986 *Appl. Opt.* **25** 2409
Geist J, Steiner B, Schaefer A R, Zalewski E F and Corrons A 1975 *Appl. Phys. Lett.* **26** 309
Goodman T M and Key P J 1988 *Metrologia* **25** 29
Martin J E, Fox N P and Key P J 1985 *Metrologia* **21** 147
Schaefer A R and Eckerle 1984 *Appl. Opt.* **23** 250
Suzuki M and Ooba N 1976 *Metrologia* **12** 123

Inst. Phys. Conf. Ser. No. 92
Paper presented at Int. Conf. Optical Radiometry, NPL, London, 12–13 April 1988

Absolute radiometric determination for the distribution of temperature of the thermal sources of light

Valentin M Feru and Gheorghe P Ispasoiu

National Institute of Metrology
Optical Radiation Measurements and Instrumentations Laboratory
11 Vitan-Birzesti Road
RO-75669, Bucharest, PO 8
Romania

This paper describes two methods referring to absolute radiometric determination for the distribution temperature of thermal light sources (Planck radiators and tungsten filament or ribbon incandescent lamps). Each of them is based on radiometric measurements in certain spectral domains using an absolute detector, and special algorithms are processing the results.

A monochromator is used in the first method and the responses of absolute detector at two different wavelengths are measured. A formula depending on spectral responsivities of absolute detector at the above wavelengths gives the distribution temperature.

In the second method the instantaneous global responses values of the absolute detector are successively determined for two appropriate different spectral domains, and two interference filters with known spectral transmission factors are used. Two different formulas are given for this second method.

The advantages and the difficulties of each methods are analysed.

The paper also gives the results of some measurements on some tungsten filament incandescent lamps originally calibrated at BIPM. The maximum different the distribution temperatures derived using our methods and the calibration values is $\sim$ 1%.

Inst. Phys. Conf. Ser. No. 92
Workshop report from Int. Conf. Optical Radiometry, NPL, London, 12–13 April 1988

195

Workshop: dissemination of radiometric scales

The workshop took the form of short presentations by an international panel, followed by general discussion from the floor.

Panel:

Dr J R Gott (chair), NPL, UK; Mr J Bastie, INM, France; Dr P Boivin, NRC, Canada; Dr A Corrons, Instituto de Optica, Spain; Dr C Fröhlich, WRC; Dr J L Gardener, CSIRO, Australia; Mr L Liedquist, SNTI, Sweden; Dr K Möstl, PTB, Fed. Rep. Germany; Mr D H Nettleton, NPL, UK; Dr J P M de Vreede, NMS, Netherlands; Dr E Zalewski, NBS, USA.

The presentations outlined the basis for realisation of radiometric scales and means of dissemination for each country. The trend is clearly towards a system of laboratory accreditation, but although a number of countries have formal arrangements for accrediting commercial calibration laboratories there are few examples to date in the field of radiometry. Several contributors outlined plans to rectify this situation, and others emphasized the need for mutual recognition of disseminated scales: international companies will experience increasing difficulties unless international equivalence of scales can be assured. Some mutual recognition agreements are already in place; others will be needed.

The discussion highlighted the difference in approach in the United States, where transfer devices and calibrated artifacts are made available to customers as the basis of dissemination of scales. This provoked a debate on the concept of 'traceability' to national standards, in particular what is meant by traceability, and whether it is feasible or desirable to limit use of the term for commercial purposes.

A discussion on measurement uncertainties drew attention to the problem of international intercomparisons where differences in measured values exceed quoted uncertainties. The discrepancies presumably arise because of unidentified systematic errors; it was felt generally that the intercomparisons were valuable in highlighting such differences so that they could subsequently be investigated further.

Other issues discussed included the requirement for stability of scales for referencing solar measurements, and the role of cryogenic radiometry both in this context and to expedite equivalence of disseminated scales internationally. It was generally agreed that the workshop had proved a successful forum for debating measurement problems, and that the ideas could be developed at future meetings.

J R Gott

National Physical Laboratory
Teddington, Middlesex TW11 OLW, UK

List of Participants

Mr S ALLCOCK
Oxford Instruments
Eynsham
OXFORD OX8 1TL

Miss V ANDERSON
National Physical Laboratory
Queens Road
Teddington
Middlesex TW11 OLW

J BASTIE
INM-CNAM
292 Rue St-Martin
75141
Paris
France

Mr J BEDFORD
Lamda Photometrics Ltd
Batford Mill
Harpenden
Herts AL5 5BZ

Dr W R BLEVIN
CSIRO
National Measurement Laboratory
PO Box 218
Lindfield
NSW 2070
Australia

Dr P BLOEMBERGEN
Dutch Metrology Service
BO Box 654
2600 AR Delft
THE NETHERLANDS

Dr L P BOIVIN
NRC
OTTAWA
Canada

Mr J BONHOURE
BIPM
Pavillon de Breteuil
F-92312
Sevres Cedex
FRANCE

Mr van den BOOM
Ned Philips Bedr BV
Postbox 80000
5600 JA Eindhoven
NETHERLANDS

Dr J CAMPOS
Instituto de Optica
SERRANO 121
28006 MADRID
SPAIN

Dr A CORRONS
Instituto de Optica
SERRANO 121
28006 MADRID
SPAIN

Dr D CROMMELYNCK
Royal Meteorological Institute
3 Avenue Circulaire
B-1180
Brussels
Belgium

Mr B DANCE
Editor
Laser Focus
8 Birmingham Rd
Alcester
Worcs. B49 5ES

Mr R DUDA
United Detector Technology
12525 Chadron Ave
Hawthorne
CALIFORNIA 90250
USA

Dr J FISCHER
PTB Institut
Abbestr. 2-12
D-1000 BERLIN 10
W GERMANY

Mr N FOX
National Physical Laboratory
Queens Road
Teddington
Middlesex TW11 OLW

Dr G H C FREEMAN
National Physical Laboratory
Queens Road
Teddington
Middlesex TW11 OLW

Dr C FROEHLICH
P.M. OBSERVATORY
World Radiation Centre
POB 173 CH-7260 DAVOS DORF
SWITZERLAND

Dr J L GARDNER
CSIRO
National Measurement Laboratory
P O Box 218
Lindfield
NSW 2070
Australia

Mr J GEIST
National Bureau of Standards
Gaithersburg
MD 20899
USA

Mr R GLOVER
Amersham International
Pollards Wood
White Lion Rd
Amersham HP7 9LL

Miss T M GOODMAN
National Physical Laboratory
Queens Road
Teddington
Middlesex TW11 OLW

Dr J R GOTT
National Physical Laboratory
Queens Road
Teddington
Middlesex TW11 OLW

Dr B GUENTHER
Standards & Calibration Office
NASA/Goddard Space Centre
Code 673
Greenbelt
Md 20771 451

L J L HAENEN
Ned Philips Bedr BV
Postbox 80000
5600 JA Eindhoven
NETHERLANDS

Dr HANIA
BNM
101 Rue de Grenelle
75700 PARIS
FRANCE

Dr R M HERRERO
Universidad Complutense
Dept de Optica
28040-MADRID
SPAIN

Mr J W HOFFMAN
Space Instruments Inc
4403 Manchester Blvd 203
Encinitas CA 92024
USA

M Van HOOSIER
Naval Research Laboratory
Code 4165
Washington
DC 20375
USA

Mr G P HURLEY
Laser Instrumentation
Unit 1
Bear Court
Daneshill
Basingstoke, Hants. RG24 OQT

Mr B JEAN
INM - CNAM
292 Rue St Martin
75141 Paris Cedex 03
France

Mr B JEAN
LCIE
33 Avenue du Gal Leclerc
92250 Fontenay aux Roses
FRANCE

Dr R KOHLER
BIPM
Pavillon de Breteuil
F-92312
Sevres Cedex
FRANCE

Dr M KUHNE
PTB
Abbestr. 2 - 12
D-1000 Berlin 10
W Germany

Mr R B LEE III
NASA Langley Research Centre
Mail Stop 420
Hampton
Virginia 23665-5225
USA

Mr L LEIDQUIST
Swedish National Testing Institute
Box 857
S-501 15 Boras
Sweden

Dr M LEROY
CNES
18 Av. E. Belin
31055 Toulouse
France

K LESZCZYNSKI
Finnish Cent. for Rad.
    and Nucl. Safety
PO Box 268
SF-00101 Helsinki
FINLAND

H LYALL
NEI IRD Co. Ltd
Fossway
Newcastle upon Tyne
NE6 2YD

Mr J E MARTIN
National Physical Laboratory
Queens Road
Teddington
Middlesex TW11 OLW

Ms B MERCIER
INM-CNAM
292 Rue Saint-Martin
75141 PARIS
CEDEX 03
FRANCE

Dr A MIGDALL
National Bureau of Standards
Gaithersburg
MD 20899
USA

MR J R MOORE
National Physical Laboratory
Queens Road
Teddington
Middlesex TW11 OLW

Dr K MOSTL
PTB
Postfach 3345
D-3300 Braunschweig
W Germany

Mr D H NETTLETON
National Physical Laboratory
Queens Road
Teddington
Middlesex TW11 OLW

Mr R PAYNE
CENTRONIC
Centronic House
King Henry's Drive
New Addington
CROYDON
Surrey CR9 OBG

Ms M PEETERS
University of Cologne
Inst. f Geophysik und Meteorologie
Kerpener Str 13
D-5000
Koln 41
W GERMANY

Mr PELLO
BIPM
Pavillon de Breteuil
F-92312
Sevres Cedex
FRANCE

Dr T J QUINN
BIPM
Pavillon de Breteuil
F-92312
Sevres Cedex
FRANCE

Dr J RYBARK
Phillips Kommun. Industr.AG
Schanzenstr. 300
5000 Koln 80
W GERMANY

Dr J E SAUVAGEAU
NBS
Division 724,03
Boulder, CO. 80303
USA

Dr A RUSSELL SCHAEFER
Science Applications Int. Corp.
11526 Sorrento Valley Rd
San Diego
CA 92121
USA

Mr M SHEERIN
Kodak Ltd
Headstone Drive
Harrow
Middx

Prof O D D SOARES
Lab Fisica - Fac Ciencias
Praca Gomes Teixera
P - 4000 PORTO
PORTUGAL

Dr K H STEPHAN
Institut Für Extraterestrisch Physik
Munich
WEST GERMANY

Dr K STOCK
PTB
Postfach 3345
D-3300 Braunschweig
WEST GERMANY

Dr E TEGELER
PTB
Abbestr. 2 12
D 1000
Berlin 10
W GERMANY

Dr J P M de VREEDE
Netherlands Metrology Service
Schoemakerstraat 97
Delft
PO Box 654
2600 AR Delft
NETHERLANDS

MR T H WARD
National Physical Laboratory
Queens Road
Teddington
Middlesex TW11 OLW

Prof W T WELFORD
Imperial College
Prince Consort Rd
LONDON SW7 2BZ

Prof. Dr B WENDE
PTB
Institut Berlin
1 Berlin 10
Abbestrasse 2-12
W GERMANY

Dr R C WILLSON
Jet Propulsion Laboratory
MS 171-400
4800 Oak Grove Drive
PASADENA CA 91001
USA

Dr J WIRTH
University of Cologne
Inst. f Geophysik und Meteorologie
Kerpener Str 13
D-5000
Koln 41
W GERMANY

Mr A M WOOLFREY
Oxford Instruments
Eynsham
OXFORD OX8 1TL

Mr P WYCHORSKI
EASTMAN KODAK
Corporate Metrology Centre
Kodak Park Bdg. 10
Rochester
NEW YORK 14650
USA

Dr E ZALEWSKI
National Bureau of Standards
Gaithersbury
MD 20899
USA

Author Index